DIVE INTO OCEAN ANIMALS

THE MOST COLORFUL CREATURES

Discover and Learn
about ocean animals bursting with color!

Mandarinfish

One of the most colorful creatures in the ocean is the Mandarinfish. Mandarinfish are found in the warm waters of the Pacific Ocean, from Hong Kong to Australia, where they hide among coral reefs and lagoons.

Their vibrant colors are not just for show — they play an important role in protecting them from predators. This phenomenon is called **aposematic coloration**, where bright and bold colors tell other animals, Stay away — I'm toxic or taste bad!"

Mandarinfish may have bright, bold colors, but they are very shy! These little fish like to stay hidden and avoid other fish whenever they can.

Did you know that these fish do not have scales? Instead, their skin is covered with a slimy coating that tastes terrible to predators, helping to keep them safe from being eaten.

Peacock Mantis Shrimp

Despite its name, the peacock mantis shrimp is not a true shrimp, but a type of crustacean called a stomatopod. It is known for its bright colors, including blue, green, and red, which help it blend into the colorful coral reefs where it lives.

These fascinating creatures mostly live in the shallow waters of the Indian and Pacific Oceans, where they spend much of their time hunting for crabs and mollusks to eat.

Peacock mantis shrimp are known for their super-fast claws! They strike so quickly that their blows are 50 times faster than the blink of your eye. In fact, their strike is believed to be the fastest of any animal on Earth! These powerful claws help them catch food and protect themselves. They're so strong, they've even been known to crack aquarium tanks!

JELLYFISH
Did you know?

Jellyfish don't have...

Challenge: Solve crossword puzzles to test your knowledge!

Spanish Shawl Sea Slug

ACROSS

3. – Their favorite snack is tiny, plant-like animals called _____
6. – The feathery orange parts on the back of the Spanish shawl sea _____
7. – The Spanish shawl sea slug is a type of _____
8. – They like _____ places where they can hide and find their _____

DOWN

Blue Dragon

ACROSS

4 – When blue dragons eat Portuguese man of wars, they can steal their stinging cells and store them in their finger-like _____
7 – Blue dragons eat dangerous animals _____

Sea Anemones

ACROSS

4 – Unlike plants, sea anemones don't need _____ to survive.
_____ to sting and catch prey
_____ eat meat like small fish and plankton

More than 80% of the ocean remains unexplored.
Who knows what colorful creatures are yet to be discovered?

OCEAN EXPLORER'S GUIDE

Are you ready to discover the most colorful ocean animals? This book makes learning fun and engaging for everyone! Whether you're a parent, teacher, or student, you'll find exciting facts, real images, and interactive activities to enjoy.

READING FUN

The book's pages are packed with fun facts and beautiful images that capture the amazing diversity of ocean life. The engaging layout ensures that every page is both educational and entertaining.

VOCABULARY

Look out for bolded terms in the text! These are key words, and you can find their definitions in the Dive Into Definitions section of the book.

SPECIES SPOTLIGHT

Each animal gets its own section of cool stats, including habitat, diet, size, and lifespan.

INTERACTIVE ACTIVITIES

At the end of the book, test your ocean animal knowledge and memory with fun crossword puzzles!

STUNNING VISUALS

Filled with bright, colorful images that shows the beauty of ocean life.

HANDS ON LEARNING

Solve fun crossword puzzles to remember cool facts about ocean animals.

EASY TO UNDERSTAND

Perfect for younger readers, with simple explanations and bolded vocabulary words to build knowledge.

HOW TO USE THIS BOOK:

PARENTS
Read aloud and explore the animal stats together for family learning. Ask questions and discuss the amazing features of each creature.

HOMESCHOOLERS
Use this book as part of a marine biology unit or science lesson. Pair it with hands-on activities, like ocean-themed crafts or aquarium visits, for a fun-packed learning experience.

TEACHERS
Use it as a classroom resource for marine biology lessons or reading time. The vocabulary and activities make it perfect for reinforcing science concepts while keeping students engaged.

KIDS
You can explore the book in any order you like! Flip to your favorite animals, read about their cool features, and challenge yourself with the crossword puzzles at the end.

Oceans are home to the largest diversity of animals on Earth.

TABLE OF CONTENTS

TABLE OF CONTENTS

EXPLORING THE
Oceans Most
COLORFUL
CREATURES

FISH

Did you know?

The oldest fish species date back over 500 million years, making them some of the earliest animals to appear on Earth.

Mandarinfish

One of the most colorful creatures in the ocean is the Mandarinfish. Mandarinfish are found in the warm waters of the Pacific Ocean, from Hong Kong to Australia, where they hide among coral reefs and lagoons.

Their vibrant colors are not just for show — they play an important role in protecting them from predators. This phenomenon is called **aposematic coloration,** where bright and bold colors tell other animals, 'Stay away — I'm toxic or taste bad!'"

Mandarinfish may have bright, bold colors, but they are very shy! These little fish like to stay hidden and avoid other fish whenever they can.

Did you know that these fish do not have scales? Instead, their skin is covered with a slimy coating that tastes terrible to predators, helping to keep them safe from being eaten.

Species Spotlight ▶ Mandarinfish

Habitat: Shallow coral reefs and lagoons in the Pacific Ocean, especially around Southeast Asia and Australia

Diet: primarily eats small crustaceans like copepods and amphipods

Size: About 2.5 to 3 inches (6 to 8 cm) long.

Weight: Less than 1 ounce

Lifespan: Around 10 to 15 years in the wild

Other Common Names: Mandarin dragonet, striped mandarinfish, green mandarinfish.

Scientific Name: Synchiropus splendidus

Royal Gramma Fish

The royal gramma is a brightly colored fish that lives in the tropical reefs of the western Atlantic Ocean. Even though they come from natural reef environments, they are commonly kept in aquariums.

The royal gramma is generally a peaceful fish that gets along well with many other kinds of fish. However, it is very protective of its territory. Even though they're small, they're brave—royal grammas can scare away larger fish by opening their mouths wide as a warning.

Check out the royal gramma's black spot on its dorsal fin! This unique feature adds to the fish's stunning appearance.

The Royal Gramma sometimes acts as a helpful "cleaner fish", picking off parasites that live on the skin of other fish.

Habitat: Coral reefs of the tropical western Atlantic Ocean, including the Caribbean Sea and the Gulf of Mexico

Diet: These fish are **planktivorous**. Their diet consists mostly of zooplankton, crustaceans and parasites.

Size: About 3 inches

Weight: Less than 0.1 ounces

Lifespan: 5- 6 years

Other Common Names: Fairy Basslet

Royal Dottyback Fish

Don't get the Royal Gramma and Royal Dottyback mixed up—they may look alike, but they're two very different fish! Check out the picture below to spot the differences.

No black spot on dorsal fin

Translucent fins

The Royal Dottyback has a more sharply divided color pattern with a clear line between the purple and yellow.

Blue eyes

Smaller mouth

Temper: meanie, aggressive

13

Butterflyfish

There are about 129 different species of butterflyfish. They are small, brightly colored fish that live in warm ocean waters around coral reefs. They have thin, flat bodies with bold patterns in colors like yellow, white, black, and orange.

Some butterflyfish species have "false eye" spots near their tails to confuse predators. These spots make it look like the butterflyfish's head is on the opposite end of its body. This helps the fish escape by quickly swimming away headfirst when threatened!

Bluecheek Butterflyfish

Wrought Iron Butterflyfish

Butterflyfish are known for forming strong bonds, often swimming in pairs throughout their lives to find food and stay safe. If they get separated, one might swim upward like a signal to say, "Hey, I'm over here!"

Raccoon Butterflyfish

One of its coolest tricks is its ability to change colors! At night, its bright colors fade to help it blend in with the reef and stay safe. But if it feels threatened, its colors can glow even brighter to scare off enemies.

Species Spotlight Butterflyfish

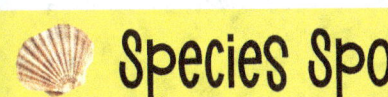

Habitat: Coral reefs and rocky areas in tropical and subtropical oceans, primarily in the Atlantic, Indian, and Pacific Oceans.

Diet: These fish are **Omnivores**. Their diet consists of coral polyps, algae, plankton, worms, and crustaceans.

Size: Most species range from 4 to 8 inches (10 to 20 cm) in length

Weight: Usually between 1 to 5 ounces

Lifespan: 5- 10 years

Scientific Name: Chaetodontidae

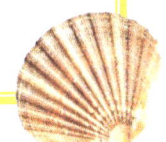

Copperband Butterflyfish

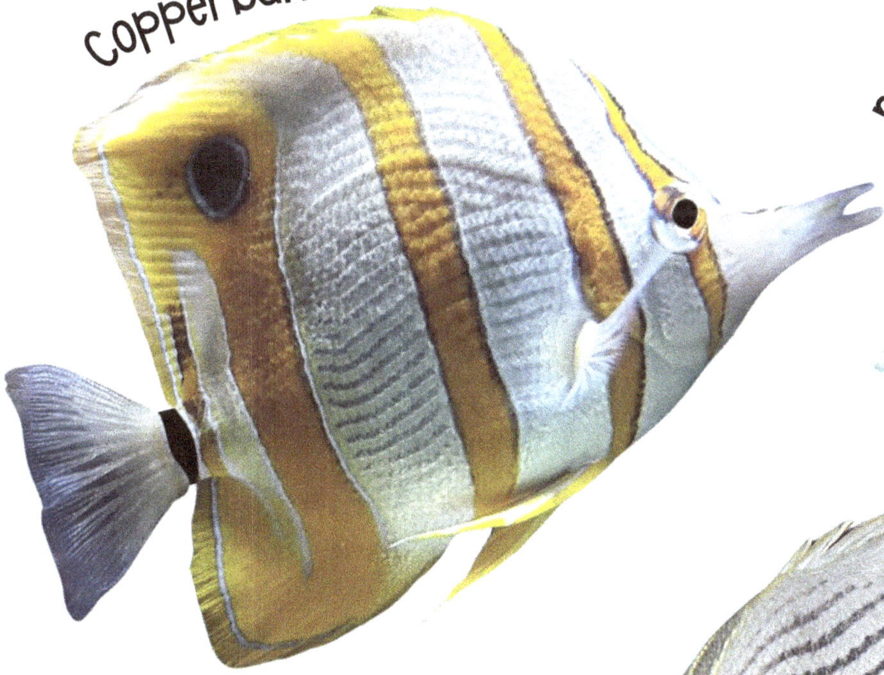

Double-Saddle Butterflyfish

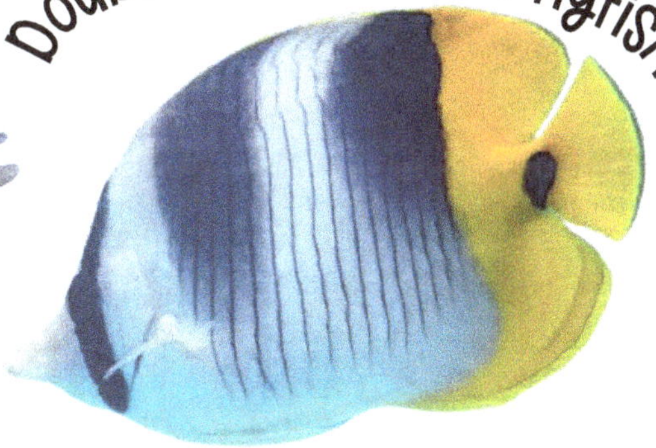

Four-Eye Butterflyfish

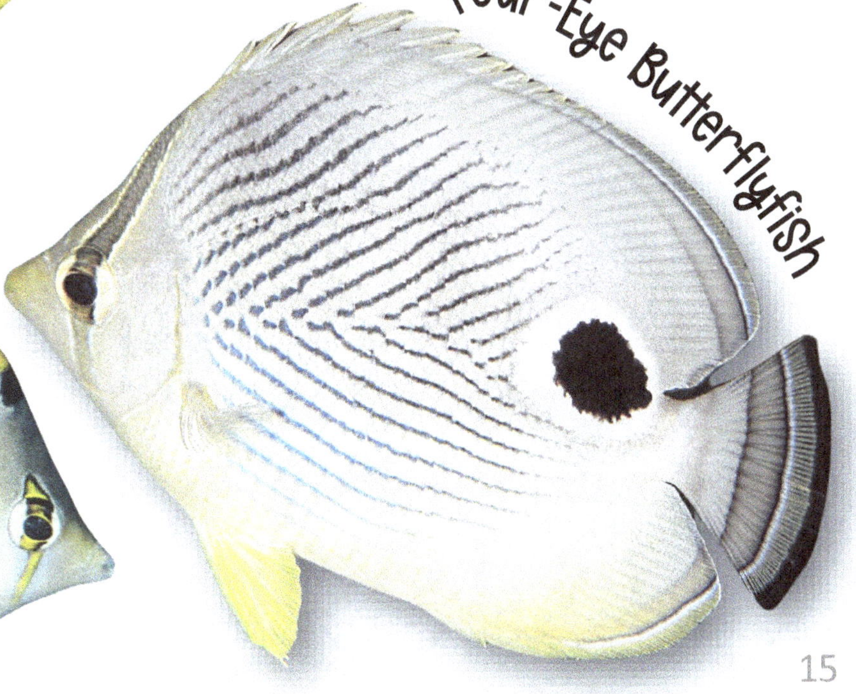

Spotband Butterflyfish

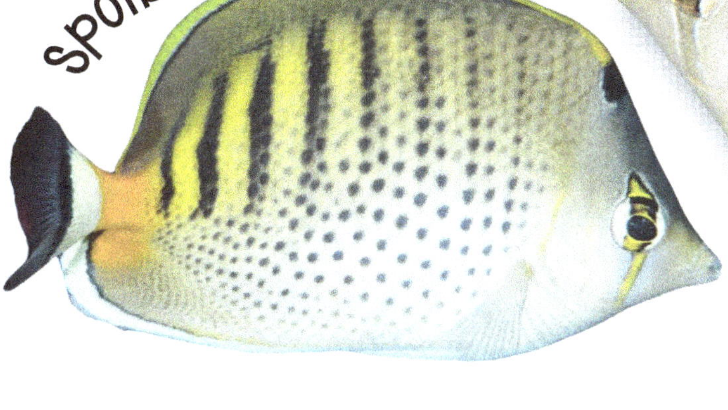

Lionfish

The lionfish's bold stripes and long spines are like a warning sign saying, 'I may look pretty, but I'm tough to mess with!' Lionfish are mostly active at night, making them nocturnal hunters. During the day, they often hide among rocks or coral, coming out after sunset to search for food.

Lionfish have 18 venomous spines that protect them from predators. While its sting isn't deadly to humans, it can cause a lot of pain.

Lionfish are not picky eaters. They gobble up small fish, shrimp, and other creatures that they can fit in their mouths. Their bodies are made to take in as much food as possible.

Their stomachs can expand up to 30 times their normal size, and their mouths can expand to more than half of their own body length!

Species Spotlight > Lionfish

Habitat: Coral reefs, rocky areas, and lagoons, mostly in the Pacific and Indian Oceans (but now in the Atlantic too!)

Diet: Small fish, shrimp, and other tiny sea creatures that can fit in its mouth

Size: About 12–15 inches long

Weight: Up to 2.6 pounds

Lifespan: 10–15 years

Other Common Names: scorpion fish, turkey fish, dragon fish, firefish, and zebrafish

Lionfish quietly spread their fins like a net to trap their prey before striking with lightning speed.

Clownfish

Clownfish are some of the most recognizable fish in the ocean. They became superstars after starring in the Disney Pixar movie Finding Nemo! But what makes clownfish extra special is their unique friendship with sea anemones!

Sea anemones are ocean animals that look like underwater flowers, but they have stinging tentacles to catch prey. Clownfish have mucus (a protective slime) on their skin that allows them to live safely among the anemone's tentacles. In return, the clownfish help keep the anemone clean by eating food scraps and parasites. They also wiggle around to keep water moving. This special teamwork is called **a symbiotic relationship**—both the clownfish and the anemone help each other survive!

Because of their special relationship with sea anemones, Clownfish are also known as Clown Anemonefish.

How to pronounce Anemone: (uh–NEM–uh–nee) Try saying it 3 times fast!

Species Spotlight   Clownfish

Habitat: Coral reefs and lagoons in the warm, tropical waters of the Indian and Pacific Oceans, including the Great Barrier Reef.

Diet: Algae, plankton, and small bits of food that fall from their anemone or are left behind by other creatures

Size: 3 - 5 inches

Weight: Less than 1 ounce

Lifespan: Up to 10 years in the wild

Other Common Names: Anemonefish

Angelfish

Angelfish are some of the most colorful and beautiful fish in the ocean! There are around 85 different species of angelfish, and they come in a rainbow of colors, including blue, yellow, orange, and even stripes or spots. These fish have flat, disk-shaped bodies and long, flowing fins.

One of the coolest things about angelfish is that their colors can change as they grow up! For example, a young angelfish might have bold black stripes that fade into bright yellow as it gets older. Healthy angelfish tend to have bright, vibrant colors, while stress, illness, or poor conditions can cause their colors to fade or become dull. Changes in diet, water quality, or the presence of predators can also impact their color.

French Angelfish

Emperor Angelfish

Flame Angelfish

Queen Angelfish

Rock Beauty Angelfish

Regal Angelfish

Yellowbar Angelfish

Lemonpeel Angelfish

 Species Spotlight   Angelfish

Habitat: Coral reefs and rocky areas in tropical and subtropical oceans around the world.

Diet: Sponges, algae, and tiny sea creatures like plankton (varies by species).

Size: Most angelfish are between 6 to 12 inches, but some species can grow as small as 2.5 inches or as large as 24 inches!.

Weight: Most angelfish weigh just a few ounces, but the biggest species can weigh up to 4 pounds!

Lifespan: 10 to 15 years in the wild.

Other Common Names: Marine angelfish

Pygmy Seahorse

Did you know that pygmy seahorses are some of the smallest seahorses in the ocean? There are eight known species of pygmy seahorses, and Satomi's pygmy seahorse is the tiniest of them all. It's only about half an inch long—just as small as a grain of rice! It's so small, it could even fit on your fingernail!

Pygmy seahorses are fish, just like clownfish and angelfish, even though they don't look like most fish. Unlike most fish, seahorses swim upright and are very slow swimmers.
They are also experts at hiding, blending perfectly into the corals where they live. Their bodies match the color and bumpy texture of coral, making them almost impossible to see.

Fun Fact: Pygmy seahorses were discovered by accident when scientists noticed them blending into the coral they were studying!

Bargibant's Pygmy Seahorse

22

The World's Tiniest Seahorses

Satomi's Pygmy Seahorse
Discovered in 2008, it is considered one of the smallest known seahorses.

Denise's Pygmy Seahorse

Identified in 2003, Denise's Pygmy Seahorse is one of the smallest seahorses, slightly larger than Satomi's Pygmy Seahorse.

Japanese Pygmy Seahorse

Identified in 2018, this species is native to the waters around Japan.

Species Spotlight — Pygmy Seahorse

Habitat: Coral reefs and sea fans in tropical waters of the Indo-Pacific

Diet: Tiny plankton and small shrimp-like creatures

Size: 0.5 to 1 inch

Weight: Less than 1 gram

Lifespan: 1 to 3 years

Scientific Name: Hippocampus Bargibanti

SHRIMP

Did you know?

Shrimp have blue blood! That's because their blood has something called hemocyanin in it. Hemocyanin helps carry oxygen through their bodies, just like how our blood uses hemoglobin to carry oxygen. But while hemoglobin makes our blood red, hemocyanin turns shrimp blood blue when it mixes with oxygen! Cool, right?

and SLUGS

Did you know?

Sea slugs produce their own slippery slime, which helps them glide smoothly over rough surfaces like coral or even sharp rocks without getting hurt.

Harlequin Shrimp

Found in the warm waters of the Indian and Pacific Oceans, harlequin shrimp are known for their unique behavior and striking appearance. These shrimp are tiny but fierce predators that feast on a very specific meal—sea stars!

Using their strong, flat pincers, harlequin shrimp work together to capture sea stars many times their size. It can take them anywhere from several days to as long as two weeks to finish eating a sea star. They often work in pairs, with one shrimp acting as a lookout while the other eats. The shrimp flip the sea star over and slowly eat it bit by bit, starting with the soft underside.

Harlequin Shrimp eating a sea star (starfish)

Harlequin shrimp are shy and calm. They like to hide during the day and come out at night to eat. They move slowly and gracefully, almost like they're dancing, while waving their claws and antennae.

They have long, thin **antennae** that help them feel and sense motion in the water. They also have shorter, petal-like **antennules**, which they use to smell and find food.

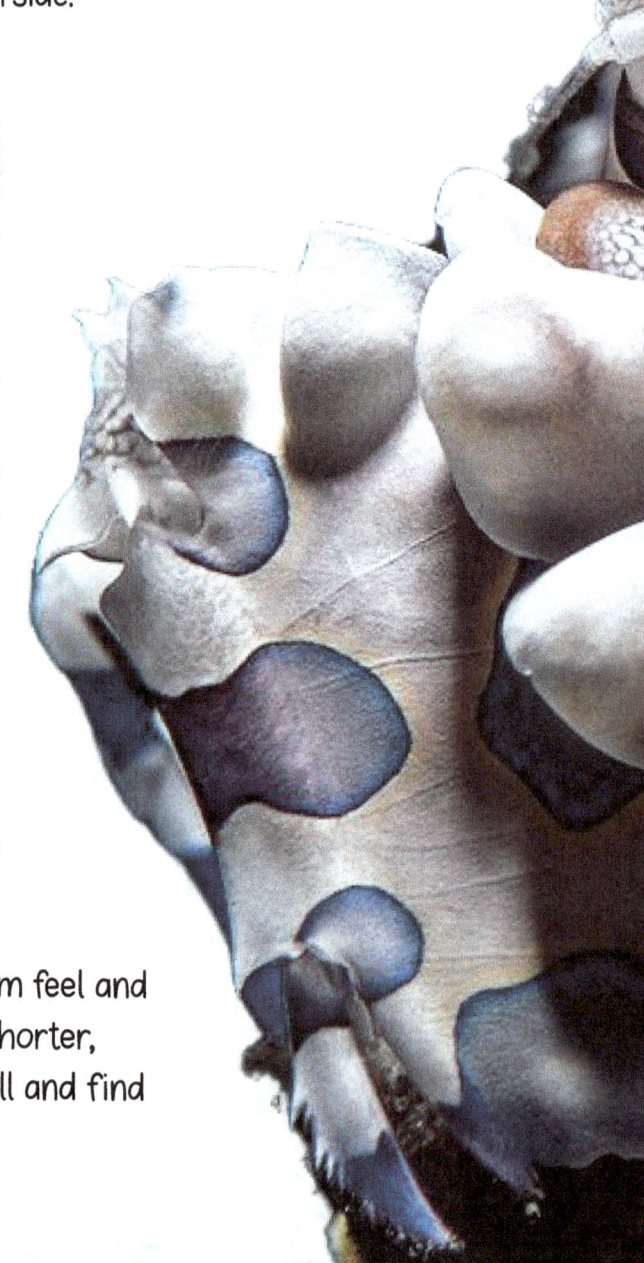

Species Spotlight ▸ Harlequin Shrimp

Habitat: Warm, tropical waters of the Indian and Pacific Oceans. They inhabit coral reefs and rubble areas, often found in sheltered environments such as crevices, caves, and under rocks.

Diet: Sea Stars

Size: About 1 to 2 inches long (as small as a paperclip!)

Weight: a few ounces

Lifespan: up to 7 years

Other Common Names: Painted Shrimp

Peacock Mantis Shrimp

Despite its name, the peacock mantis shrimp is not a true shrimp, but a type of crustacean called a stomatopod. It is known for its bright colors, including blue, green, and red, which help it blend into the colorful coral reefs where it lives.

These fascinating creatures mostly live in the shallow waters of the Indian and Pacific Oceans, where they spend much of their time hunting for crabs and mollusks to eat.

Peacock mantis shrimp are known for their super-fast claws! They strike so quickly that their blows are 50 times faster than the blink of your eye. In fact, their strike is believed to be the fastest of any animal on Earth! These powerful claws help them catch food and protect themselves. They're so strong, they've even been known to crack aquarium tanks!

Peacock mantis shrimp have some of the most incredible eyes in the ocean! Their eyes are like tiny, powerful cameras that can see colors we can't even imagine, including ultraviolet light. Each eye can move on its own, so they can look in two different directions at the same time!

Did you know? Ultraviolet (UV) light is a type of light that comes from the sun, but we can't see it with our eyes because it's invisible to humans.

Species Spotlight — Peacock Mantis Shrimp

Habitat: Shallow parts of the Indian and Pacific Oceans, near coral reefs

Diet: Small fish, mollusks, crabs and other crustaceans

Size: 2 to 7 inches

Weight: 0.4–3.2 ounces (12–90 grams)

Lifespan: 3 to 6 years, with some living up to 20 years

Other common names: harlequin mantis shrimp, painted mantis shrimp, clown mantis shrimp, rainbow mantis shrimp

Spanish Shawl Sea Slug

The Spanish shawl sea slug is bright purple with feathery orange parts on its back called cerata. It also has red antennae-like structures on its head, called rhinophores, which help it smell and sense its surroundings. Its bright colors aren't just for show—they help warn predators to stay away because it might taste bad or be toxic.

The Spanish shawl sea slug, like many other sea slugs, has a body made mostly of water. This gives it a soft, squishy, and jelly-like feel. What makes the Spanish shawl even more special is the way it moves. If it feels scared, it can swim by flipping its body back and forth like it's doing a little underwater dance!

These sea slugs are tiny, growing up to about 3 inches long, which is about the size of a crayon. They live in shallow waters along the west coast of North America, from Canada to Mexico. They like rocky places where they can hide and find their favorite snack—tiny, plant-like animals called hydroids.

Rhinophores

The red, antennae-like structures on its head help it detect smells and sense its surroundings.

Their jelly-like bodies are primarily made of water.

Rhinophore Tentacles

used for sensing the environment

The Spanish shawl is a type of nudibranch. But what is a nudibranch? It's a soft, colorful sea slug that lives in the ocean and is famous for its incredible patterns and shapes.

cerata

These soft, finger-like structures help the slug breathe by increasing its surface area, making it easier to absorb oxygen from the water.

🐚 Species Spotlight   Spanish Shawl Sea Slug

Habitat: Found in the warm waters of the Pacific Ocean, particularly around California, Mexico, and Baja California. They live in rocky areas and coral reefs, often in shallow waters.

Diet: Feeds on hydroids, small marine invertebrates, and sometimes sponges.

Size: Up to 3 inches long.

Weight: Approximately 0.05 ounces (1.4 grams).

Lifespan: Short — from a few months to a year

Scientific Name:: Flabellina iodinea

Leaf Sheep

The leaf sheep sea slug looks like a cartoon sheep covered in leaves. Its name comes from its leafy, green cerata (the small, plant-like structures on its back) that make it look like a living plant. But here's the coolest part: the leaf sheep doesn't just look like a plant—it acts like one too!

The leaf sheep can perform **kleptoplasty**, a process where it steals chloroplasts (the parts of a plant cell that make energy from sunlight) from the algae it eats. Once the chloroplasts are inside its body, the leaf sheep can use sunlight to create energy, just like a plant!

The leaf sheep lives in shallow tropical waters on algae-covered coral reefs. You can find it in the Indo-Pacific region, near places like Japan, the Philippines, and Indonesia. It spends its days happily munching on algae!

Habitat: Found in shallow, tropical waters of the Indo-Pacific, often near Japan, the Philippines, and Indonesia. It lives on algae-covered coral reefs.

Diet: Feeds on algae, with a preference for specific types like Avrainvillea.

Size: About 0.2 to 0.4 inches long. (It is so tiny that it can easily fit on your fingernail!)

Weight: Less than 0.01 ounces

Lifespan: Around 6 months to 1 year

Scientific Name:: Costasiella kuroshimae

The leaf sheep is often called the "solar-powered sea slug".

Blue Dragon

Have you ever seen a dragon swimming in the ocean? While you won't find any fire-breathing monsters in the sea, you might spot something just as magical — the blue dragon sea slug!

The blue dragon's body is silvery-white and blue, with six arms that spread out like wings. These "wings" are actually groups of finger-like parts called cerata that help it float upside down on the ocean's surface! Their blue side faces upward to blend with the water, while their silvery side faces downward to camouflage with the sky, helping them avoid predators.

Blue dragons eat dangerous animals called Portuguese man-of-war (which look like jellyfish). When they eat these creatures, they can steal their stinging cells and store them in their finger-like cerata. This means the blue dragon can now use these stings to defend itself! It's like taking someone else's superpower and making it your own.

Portuguese Man of War

Blue dragons live in warm oceans around the world. While they usually stay out in the open ocean, strong winds and currents sometimes wash them onto beaches. If you ever see one on the beach, remember: never touch it! Even though they're beautiful, those stolen stinging cells can still hurt you.

Species Spotlight — Blue Dragon

Habitat: Warm waters of the Atlantic, Pacific, and Indian Oceans; floats on the ocean's surface

Diet: Portuguese man-of-war (their favorite), blue buttons (a small marine organism), other small venomous sea creatures

Size: 1..2 inches

Weight: 0.1 and 3.5 ounces

Lifespan: about 1 year

Scientific Name: Glaucus atlanticus

OCTOPUS

Did you know?

Did you know that all octopuses have three hearts and nine brains? They have one big brain in the center, right between their eyes, and a mini brain in each of their eight arms! How cool is that?

JELLYFISH
Did you know?

Jellyfish don't have brains, hearts, or bones!

Mimic Octopus

Have you ever heard of an animal that can pretend to be other animals? Meet the mimic octopus!

This incredible creature doesn't just change colors like other octopuses — it can actually make itself look and act like at least 15 different sea animals. It can flatten its body and arms to look like a flatfish, stretch its arms out to copy a lionfish, or bury itself in the sand with just two arms showing to look like a sea snake. When it swims through open water, it can even make itself look like a jellyfish by puffing up its head and letting its arms trail behind.

These octopuses are smart! It has to remember how different animals move and act, and then copy them perfectly. What makes this octopus even more incredible is that it can switch between these disguises in seconds — even while its moving across the ocean floor! This special skill helps it stay safe from predators and catch food more easily.

These amazing creatures live in the warm waters near Indonesia, Malaysia, and other parts of Southeast Asia. They prefer to live in places with sandy or muddy bottoms where they can easily hide and hunt for food. Unlike many octopuses that come out at night, mimic octopuses are active during the day.

Species Spotlight

Mimic Octopus

Habitat: Muddy sea floors in shallow tropical waters of the Indo-Pacific, especially near Indonesia and Malaysia.

Diet: Small fish, crabs, and worms

Size: Up to 2 feet long, including arms

Weight: About 1-2 pounds

Lifespan: About 9 months to 2 years

Scientific Name: Thaumoctopus mimicus

Blue-Ringed Octopus

The blue-ringed octopus might be small —about the size of a golf ball— but it's one of the most dangerous creatures in the ocean! When it's calm, this little octopus looks pretty plain, with a yellowish-brown or beige body. But when it feels threatened, something amazing happens: bright blue rings glow all over its body in less than a second! These glowing rings are a clear warning to predators (and people!) to stay far away.

Did you Know? The blue-ringed octopus has some of the most powerful venom in all the oceans. It can produce enough venom to harm 26 humans at once! That's more powerful than many animals hundreds of times its size!

Like other octopuses, the blue-ringed octopus is an expert at hiding. With no bones, it can squeeze into tiny spaces in coral reefs and rocks. It also has the ability to change its color and texture to blend perfectly with its surroundings. This helps it stay safe from predators and surprise the small creatures it hunts for food.

Species Spotlight ▶ Blue-Ringed Octopus

Habitat: Tide pools and coral reefs in the Pacific and Indian Oceans, including Australia and Japan.

Diet: Crabs, shrimp, and small fish.

Size: : 5 to 8 inches (12-20 cm) including arms.

Weight: Less than 1 ounce (about the weight of a coin).

Lifespan: About 2 years

Scientific Name: Hapalochlaena

Mauve Stinger Jellyfish

The mauve stinger jellyfish, or Pelagia noctiluca, is a small but amazing ocean creature that is bioluminescent, meaning it can glow in the dark! It does this by making a glowing slime from its skin.

Their bodies also have tiny red spots filled with stinging cells to help them catch food and stay safe. While the sting can hurt, it's usually not dangerous to humans.

Did you Know? Their scientific name "noctiluca" means "night light" in Latin.

The mauve stinger jellyfish often travels in massive groups called blooms.

Species Spotlight 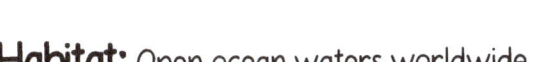 Mauve Stinger Jellyfish

Habitat: Open ocean waters worldwide.

Diet: Zooplankton, including small crustaceans

Size: : Bell diameter of 3–12 cm; tentacles up to 3 meters long.

Weight: Under .05 kg (.11 lb)

Lifespan: 6 – 9 months

Scientific Name: Pelagia noctiluca

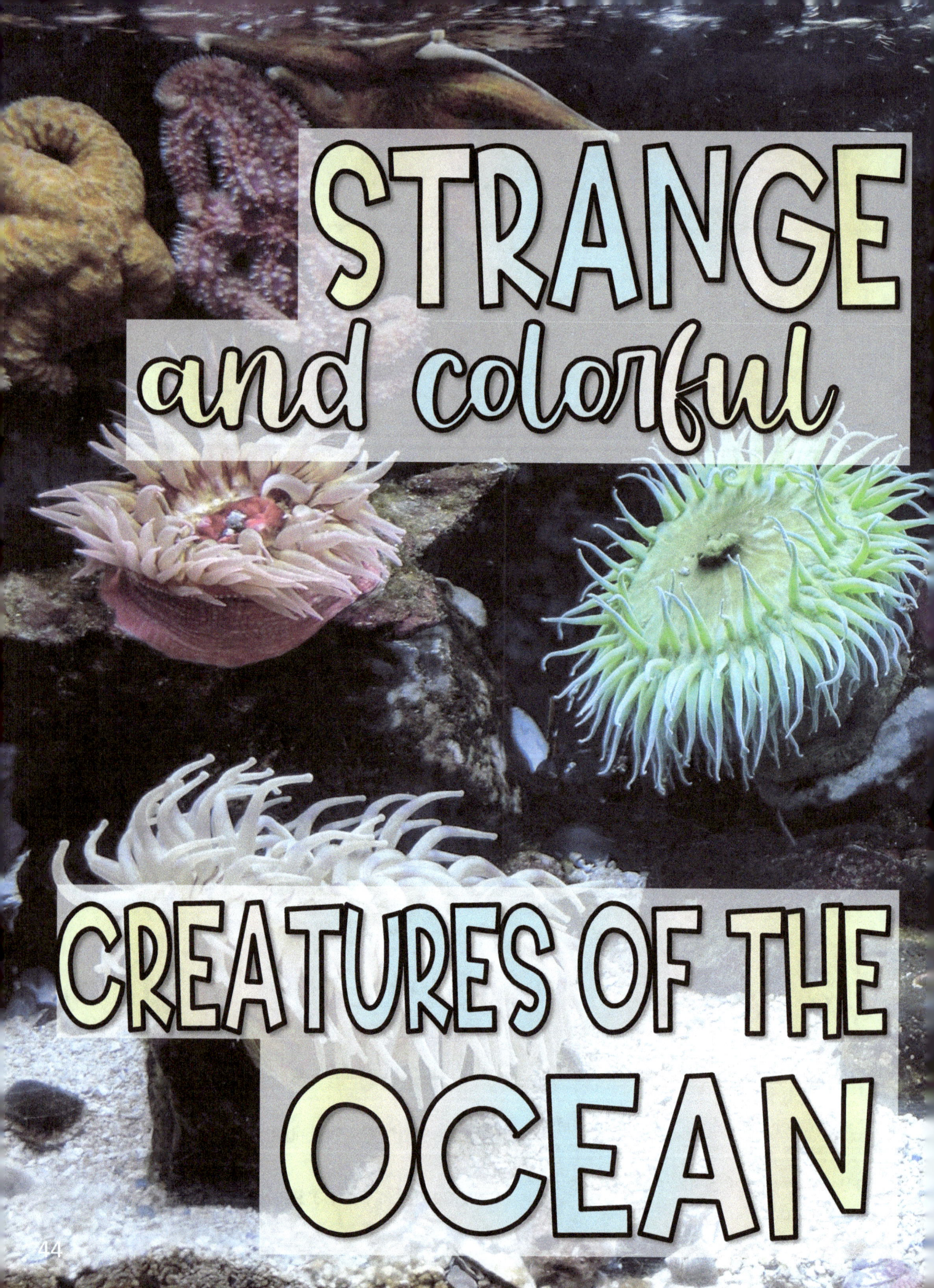

STRANGE and colorful

CREATURES OF THE OCEAN

Did you know?

From shallow coral reefs to the freezing depths of Antarctica, sea anemones live in oceans all over the world.

Leafy Seadragon

Leafy seadragons live in the shallow coastal waters of southern Australia, spending their days gently drifting through seagrass beds and kelp forests. These marine fish belong to the same family as seahorses and pipefish, but they don't swim like typical fish. Instead, they float around like a piece of seaweed and use tiny, see-through fins near their head to move.

Leafy seadragons are masters of disguise. They look just like a piece of seaweed, with leaf-like parts sticking out all over their bodies. These "leaves" aren't plants—they're part of the fish's body and help it blend in to avoid predators and sneak up on prey.

They live in some of the most vibrant underwater landscapes—kelp forests, seagrass meadows, and rocky reefs, which provide them with both food and shelter. Unfortunately, their homes are under threat from pollution and human activities, which is why leafy seadragons are now protected by Australian law.

Leafy seadragons feed on tiny creatures like mysid shrimp and plankton. Instead of teeth, they have a long, tube-like snout they use to suck up prey—like sipping through a straw! One surprising fact is that the leafy sea dragon has no stomach, so it must eat frequently to keep its energy up.

Species Spotlight

Leafy Seadragon

Habitat: Southern and eastern Australian temperate waters

Diet: small crustaceans, plankton, shrimp

Size: : 20-24 centimeters (8-9.5 inches)

Weight: 25-30 grams (0.88-1.06 ounces)

Lifespan: 7 – 10 years

Scientific Name: Phycodurus eques

Christmas Tree Worms

Christmas tree worms are bright and colorful marine worms with spiraling shapes that look like little Christmas trees. They come in many colors, like red, orange, yellow, blue, and white — sometimes even multiple colors on the same worm. They live in warm, tropical coral reefs in oceans around the world.

Even though they look like trees, these worms are very small. They usually grow to about 1.5 inches long, with their tree-like crowns making them look bigger than they really are.

These worms use their feathery crowns to catch tiny food particles floating in the water. They help keep the water clean and provide food for some reef fish. They play an important role in keeping coral reefs healthy. They help protect corals from invasive sea stars that can harm the reef. At the same time, these worms prevent algae from overgrowing the coral, which helps keep the reef thriving.

Species Spotlight

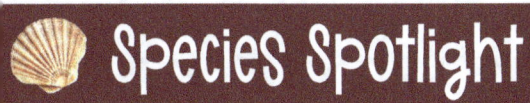

Christmas Tree Worm

Habitat: Warm, tropical coral reefs around the world

Diet: Filter feeder – eats tiny food particles floating in the water

Size: : About 1.5 inches long

Weight: very light – only a few grams

Lifespan: Around 30 years if the coral stays healthy

Scientific Name: *Spirobranchus giganteus*

When they find a spot they like, Christmas tree worms make a hard tube of calcium in the coral and stay there for life. The coral helps protect them from predators. When the worm feels safe, it opens its tree-like crowns to catch food and breathe. The crowns help the worm collect tiny food and take in oxygen. If the worm feels scared, it quickly pulls the crowns back into its tube to stay safe.

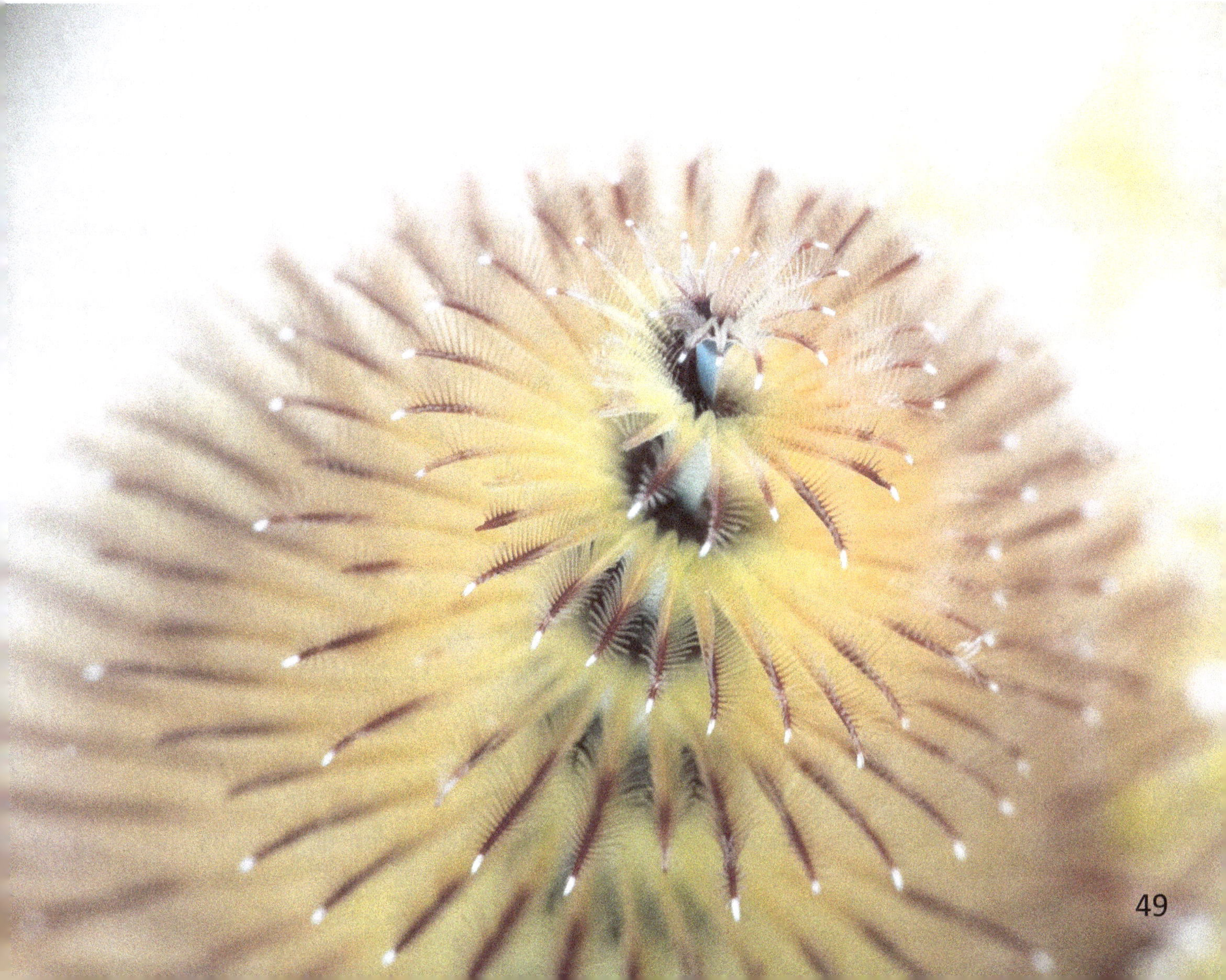

Sea Anemones

Sea anemones may look like colorful underwater flowers, but they are actually animals with soft bodies and stinging tentacles! These fascinating creatures come in many colors and shapes, brightening up coral reefs and ocean floors all over the world. Some live in shallow waters, while others thrive in the deep, dark ocean. Sea anemones are also close relatives of coral and jellyfish.

Most sea anemones are small, but some species can grow to be as wide as six feet! They use their soft, tube-shaped bodies to attach themselves to rocks, coral, or the ocean floor. Unlike plants, they don't need sunlight to survive. Instead, they catch food using their tentacles.

Sea anemones are carnivores, which means they eat meat! Their tentacles are covered with tiny stingers that help them catch small fish, shrimp, and plankton. Larger anemones have even been known to eat starfish and crabs. Once they catch their meal, they pull it into their mouth, located at the center of their body.

Species Spotlight

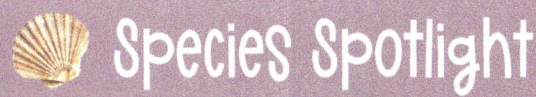

Sea Anemones

Habitat: Found in oceans all over the world, from shallow coastal waters to deep-sea environments

Diet: Carnivorous – eats small fish, plankton, and other tiny sea creatures caught with its tentacles

Size: : varies greatly – from less than 1 inch to over 6 feet wide

Weight: usually very light, but larger species can weigh several pounds

Lifespan: can live for decades, some species are believed to live over 50 years

Scientific Name: Actiniaria

These creatures might look delicate, but they are important predators in the ocean. Some sea anemones even form partnerships with clownfish, providing them with a safe home in exchange for protection.

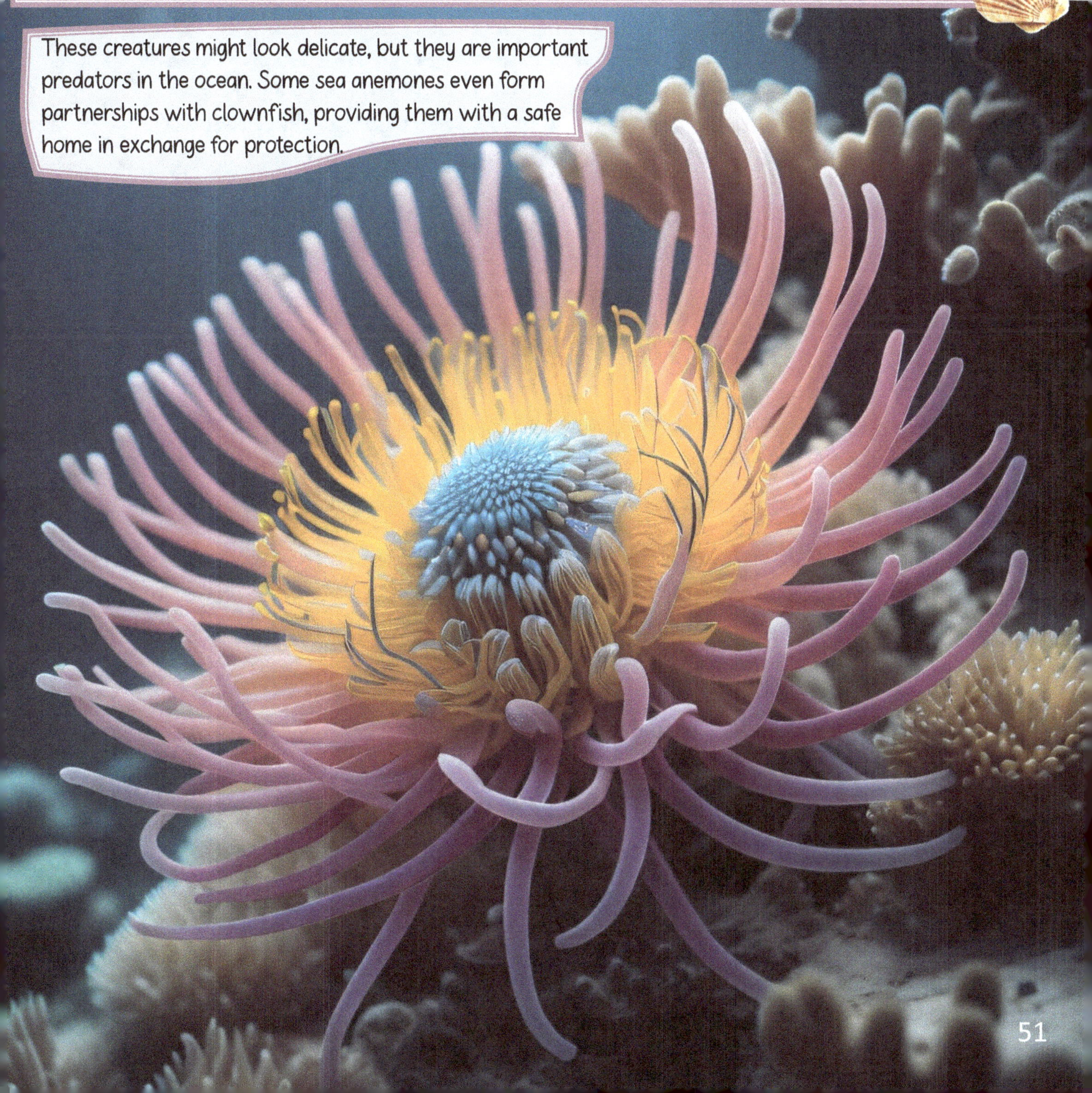

Sea Stars

Sea stars, often called starfish, are some of the most amazing creatures in the ocean! Despite their name, they aren't fish at all. Instead, they are part of a group of animals that includes sea urchins and sand dollars, which also live on the ocean floor.

Sea stars are carnivores and love to eat clams, mussels, and other shellfish. They have a very unusual way of eating: they push their stomachs out of their bodies to digest food outside before pulling it back inside!

Did you Know? Sea stars don't have a brain—or even blood! Instead, they have a simple nerve network that helps them feel and move. They also use filtered seawater to pump nutrients through their bodies and power their nervous system.

One of the coolest things about sea stars is if they lose an arm, they can grow it back over time. Some species can even grow an entirely new sea star from just one arm! However, the success of this depends on the species and how severe the injury is. The process can take several months to years.

Sea Stars

Habitat: Found in oceans all over the world, from shallow coastal waters to deep ocean floor

Diet: Carnivores; they feed on mollusks, small fish, other sea stars, and algae.

Size: : Most adult sea stars are 8 to 12 inches.

Weight: Can range from 0.1–13 lbs

Lifespan: Between 10 – 34 years

Scientific Name: Asteroidea

Sea stars come in many shapes, sizes, and colors. Most have five arms, but some species can have many more—even as many as 40! These arms are covered in tiny tube-like feet that help them move slowly along the ocean floor.

The Sunflower sea star and the Crown of thorns are examples of sea stars with more than 5 arms.

Crown of Thorns
Arms: 10 – 21

Sunflower Sea Star
Arms: 16 to 24

Sea Urchins

Sea urchins are small, round animals covered in sharp-looking spines. While they might seem plain at first, many sea urchins are surprisingly colorful. You can find them in shades of purple, red, green, pink, and even bright orange. Their spines can be long and needle-like or short and stubby.

These spiky creatures are found in every ocean on Earth. Some species make their homes near the seashore, living between the high and low tide lines. Others dwell in the deep ocean, often hiding in crevices or clinging to rocks.

Sea urchins can "see" even though they don't have eyes like we do! Instead, they use special light-sensitive spots on their tube feet. These cells allow sea urchins to sense their surroundings and navigate their environment. It's like their whole body works as one big eye! This helps sea urchins avoid predators and find shelter.

Sea urchins play an important role in keeping the ocean healthy. By eating algae, they help prevent it from overgrowing and harming coral reefs. Some animals, like sea otters and starfish, rely on sea urchins for food, making them an essential part of the underwater food chain.

Purple sea urchins

Flower urchin

Slate-pencil urchin

Species Spotlight

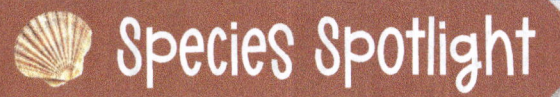

Sea Urchins

Habitat: Found in oceans all over the world

Diet: Herbivores - algae and seaweed

Size: Average size: 1-4 inches

Weight: About 1 pound

Lifespan: In the wild, depending on species, they can live anywhere from 5-100 years or longer

Scientific Name: Echinoidea

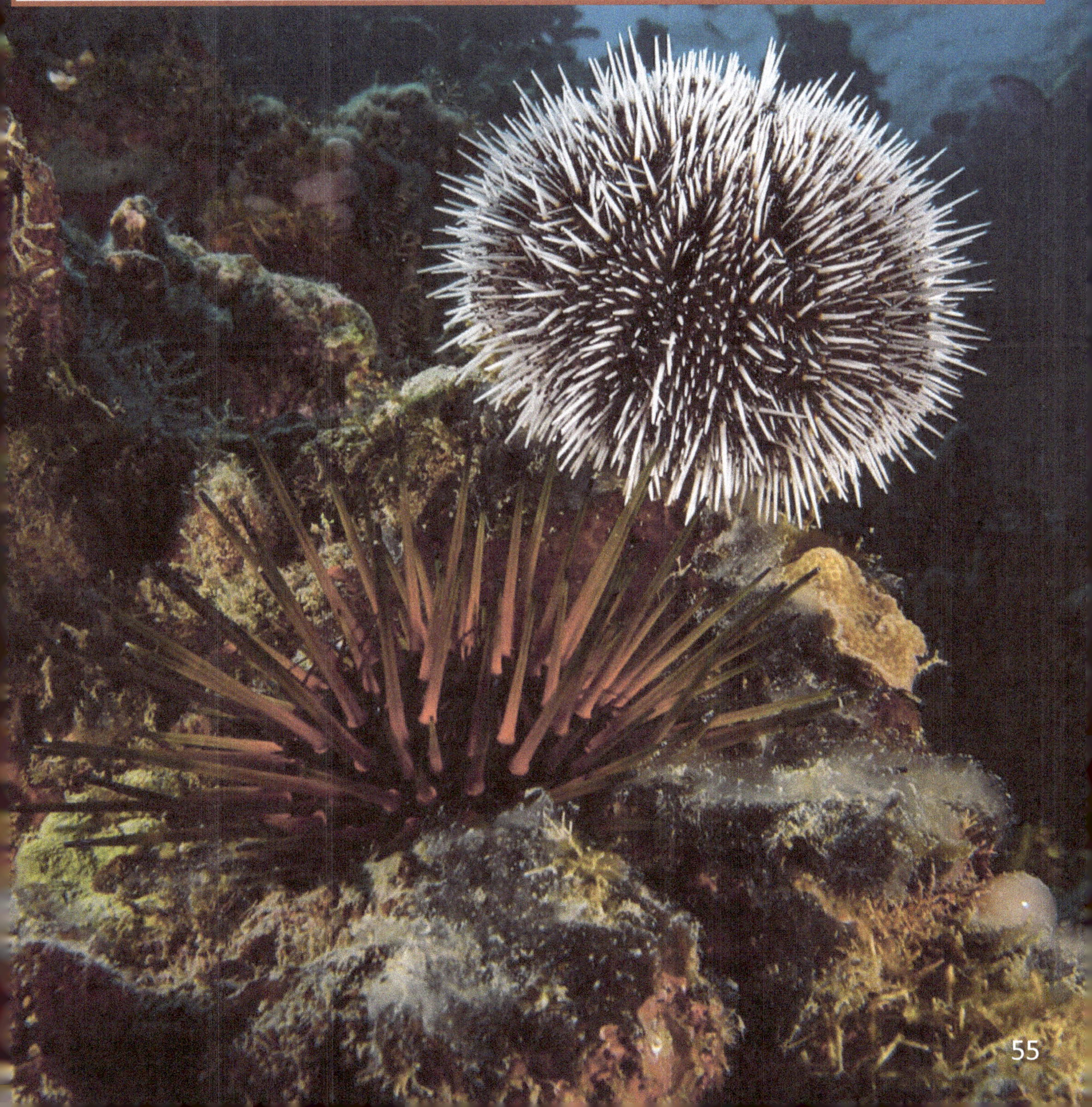

DIVE INTO DEFINITIONS

Habitat

The natural home where a plant or animal lives, with everything it needs to survive.

Camouflage

A way animals blend into their surroundings to hide from predators or sneak up on prey.

Predator

An animal that hunts and eats other animals.

Prey

An animal that is hunted and eaten by other animals.

Aposematic Coloration

Bright, bold colors or patterns on an animal that warn predators it is poisonous, venomous, or dangerous to eat.

Algae

Tiny, plant-like organisms that grow in water and can be food for ocean animals.

Plankton

Tiny plants and animals that float in the ocean and are an important food source for sea creatures.

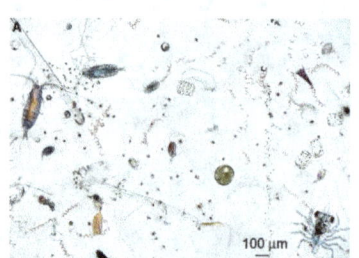

Coral Reef

Tiny plants and animals that float in the ocean and are an important food source for sea creatures.

Crustaceans

A group of animals with hard shells, like crabs, lobsters, and shrimp.

Appendages

Body parts like legs, arms, or fins that animals use to move or grab things.

Phytoplankton

Tiny plants in the ocean that make oxygen and are the base of the food chain.

Bioluminescence

The ability of some animals to glow in the dark

Adaptation

A special feature or behavior that helps an animal survive in its environment.

DIVE INTO DEFINITIONS

Ecosystem

A community of living things, like plants and animals, interacting with their environment.

Photosynthesis

The process plants and some ocean creatures use to turn sunlight into energy.

Marine

Anything related to the ocean.

Mollusk

A soft-bodied animal, often with a shell, like clams or octopuses.

Tentacles

Long, flexible body parts that some sea creatures use to feel, grab, or move.

Current

The movement of water in the ocean.

Invertebrate

An animal without a backbone, like jellyfish, crabs, and octopuses.

Gills

Body parts that fish and some other animals use to breathe underwater.

Shellfish

Sea animals with hard shells, like shrimp, crabs, and clams.

Nocturnal

Animals that are active at night and rest during the day.

Spines

Sharp, pointy parts on some animals, like sea urchins, to protect them from predators.

Venom

A poisonous substance some animals use to protect themselves or catch prey.

Lifespan

How long an animal or plant lives, from birth to death.

Aquatic

Living or growing in water.

Symbiotic Relationship

When two living things work together and help each other survive.

How Much
DO YOU REMEMBER?

Butterflyfish

ACROSS

5. – These fish can glow brighter when they feel _____ to reach food in tight spaces.
6 – The copperband butterflyfish has a long, pointed _____ reefs.
most commonly found near _____ waters around coral reefs.

SOLVE THE
Crossword Puzzles
TO FIND OUT!

Blue Dragon

Sea Urchins

ROSS

When blue dragons eat Portuguese man of wars, they can steal their stinging cells and
tore them in their finger-like _____.
lue dragons eat dangerous animals called _____.

ACROSS
4. — Sea urchins are found in every _____ on Earth.
5. — Sea urchins have sharp, spiny _____ to protect themselves.

DOWN
Sea urchins are _____, meaning they eat plants like algae.
_____ important for controlling the growth of _____ on cora

Crossword Challenge Instructions

 ## Directions:

- Read the clues carefully and fill in the crossword with the correct words.

- Use what you've learned in the book to help solve the puzzles!

 ## Pro Tips:

- If you're stuck, try looking back at the animal pages for hints.

- Challenge yourself by solving the puzzles without the word bank!

 ## Using the Word Bank:

The word bank is divided into four categories: Fish, Shrimp and Slugs, Octopus and Jellyfish, and Strange Creatures. This way, you won't have to search through one giant list!

The words are listed in no particular order, so take your time finding the right one! If you need help, check the category that matches the puzzle clue. But remember, it's even more fun to try solving without the word bank first!

If you get stuck, try flipping through the book to find the answers.

WORD BANK

Fish

Stripes	Reef	Mucus	Anemonefish
Inches	Spines	Eye	Cleaner
Hippocampus	Warm	Sponges	Camouflage
Toxic	Black	Threatened	Healthy
Fins	Ounce	Tropical	Anemone
Pacific	Chaetodontidae	Disk	Peaceful
Japanese	Night	Copepods	Fade
Pairs	Nocturnal	Lagoons	Upright
Crustaceans	Humans	Predators	Stomachs
Atlantic	Picky	Nemo	Fish
Snout	Splendidus	Basslet	Colors
Rocky	Coral	Slimy	Mouth
Algae	Satomi	Symbiotic	

Shrimp and Slugs

Kleptoplasty	Hydroids	Tiny	Photosynthesis
Antennae	Body	Atlanticus	Sun
Water	Shallow	Directions	Indian
Portuguese	Eyesight	Crabs	Cerata
Leaves	Solar	Eat	Rhinophores
Protect	Warm	Claws	Rocky
Pincers	Nudibranch	Stomatopod	Slugs
Floating	Pairs	Painted	Ultraviolet
Stars	Breathe	Algae	Costasiella

WORD BANK

Octopus and Jellyfish

Venom	Seconds	Hapalochlaena	Pelagia
Blooms	Predators	Hiding	Color
Asia	Muddy	Day	Skin
Shrimp	Flatfish	Rings	Eight
Animals	Sting	Dark	Poisonous

Strange Creatures

Tentacles	Australia	Clean	Coral
Float	Spines	Environment	Eyes
Herbivores	Symbiotic	Snout	Outside
Ocean	Fifty	Thirty	Flowers
Arms	Calcium	Healthy	Filter
Algae	Law	Stomach	Christmas
Eques	Crown	Network	Feet
Sunlight	Pipefish	Jellyfish	Actiniaria
Carnivores	Brain	Oxygen	Camouflage

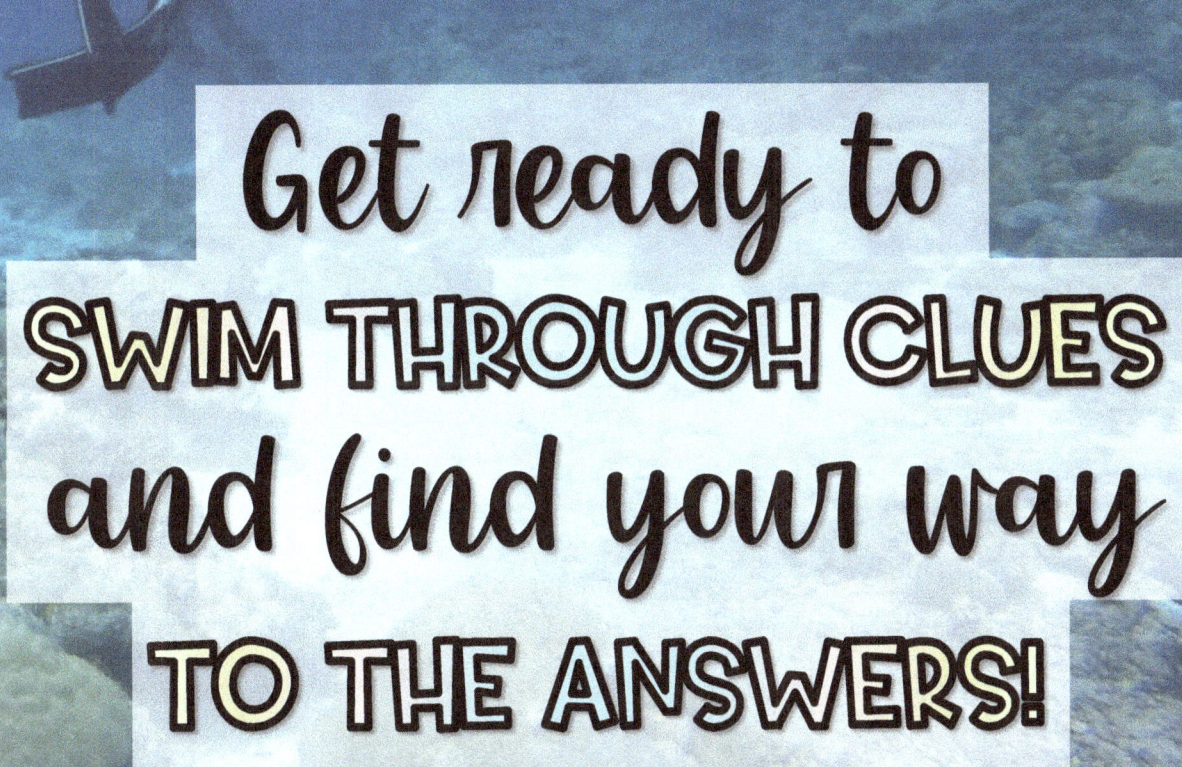

Get ready to
SWIM THROUGH CLUES
and find your way
TO THE ANSWERS!

Mandarinfish

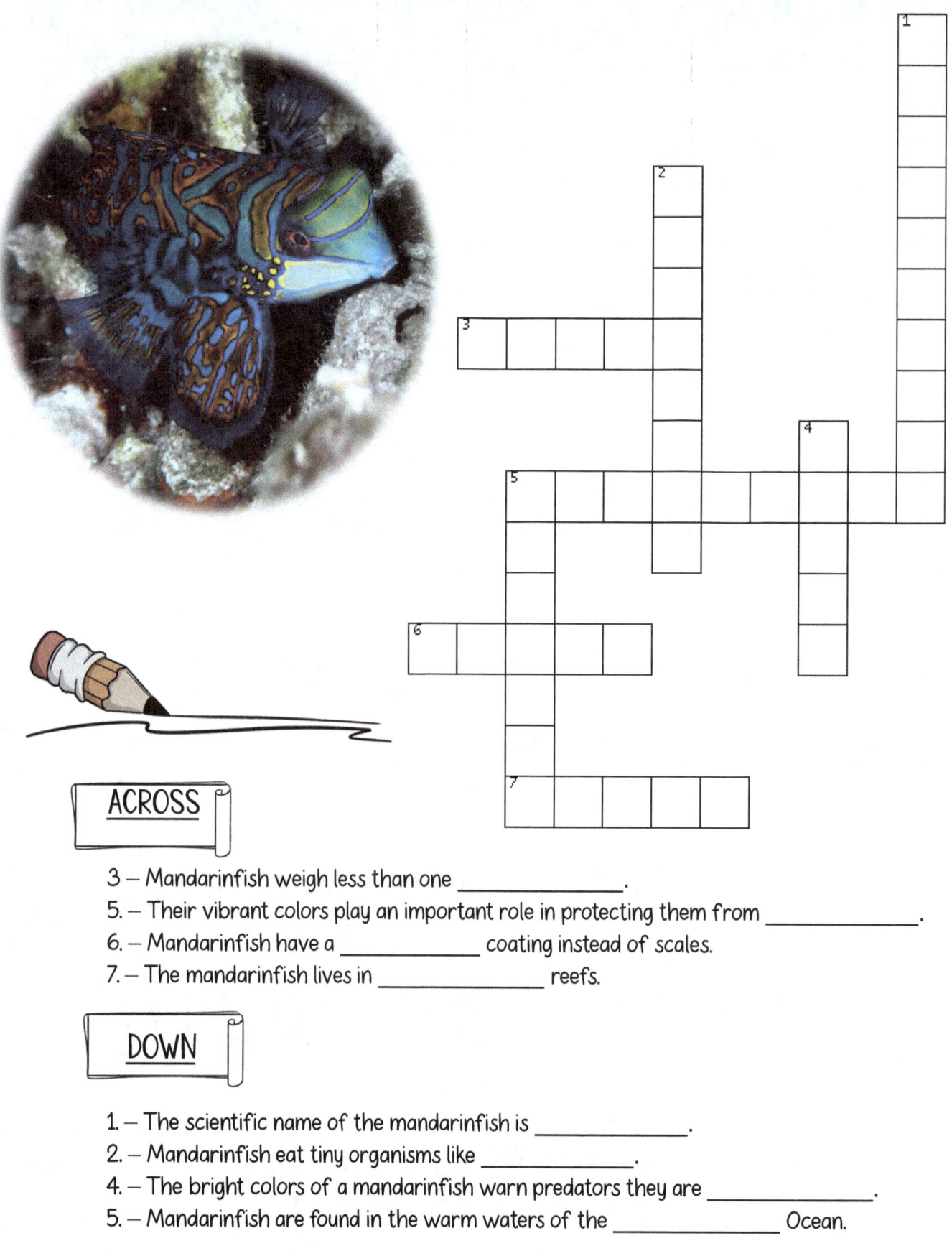

ACROSS

3 — Mandarinfish weigh less than one _____.

5. — Their vibrant colors play an important role in protecting them from _____.

6. — Mandarinfish have a _____ coating instead of scales.

7. — The mandarinfish lives in _____ reefs.

DOWN

1. — The scientific name of the mandarinfish is _____.

2. — Mandarinfish eat tiny organisms like _____.

4. — The bright colors of a mandarinfish warn predators they are _____.

5. — Mandarinfish are found in the warm waters of the _____ Ocean.

Royal Gramma Fish

ACROSS

2. — The royal gramma fish come from natural _____ environments.
4. — The royal gramma has a _____ spot on its dorsal fin.
5. — Royal grammas eat tiny creatures like plankton and _____.
7. — A royal gramma can scare away larger fish by opening its _____ wide as a warning.
8. — Royal grammas are sometimes called fairy _____.

DOWN

1. — The royal gramma is generally a _____ fish that gets along well with many other fish.
3. — The royal gramma sometimes acts as a helpful "_____" fish, picking off parasites that live
    on the skin of other fish.
6. — They are found in the western _____ Ocean.

Butterflyfish

ACROSS

5. – These fish can glow brighter when they feel _____.
6. – The copperband butterflyfish has a long, pointed _____ to reach food in tight spaces.
7. – Butterflyfish are most commonly found near _____ reefs.
8. – Butterflyfish live in _____ ocean waters around coral reefs.

DOWN

1. – Butterflyfish are often seen swimming in _____ to find food and stay safe.
2. – The scientific name for the butterflyfish family is _____.
3. – Butterflyfish have "false _____" spots near their tails to confuse predators.
4. – At night, butterflyfish colors _____ to blend in with the reef.

Lionfish

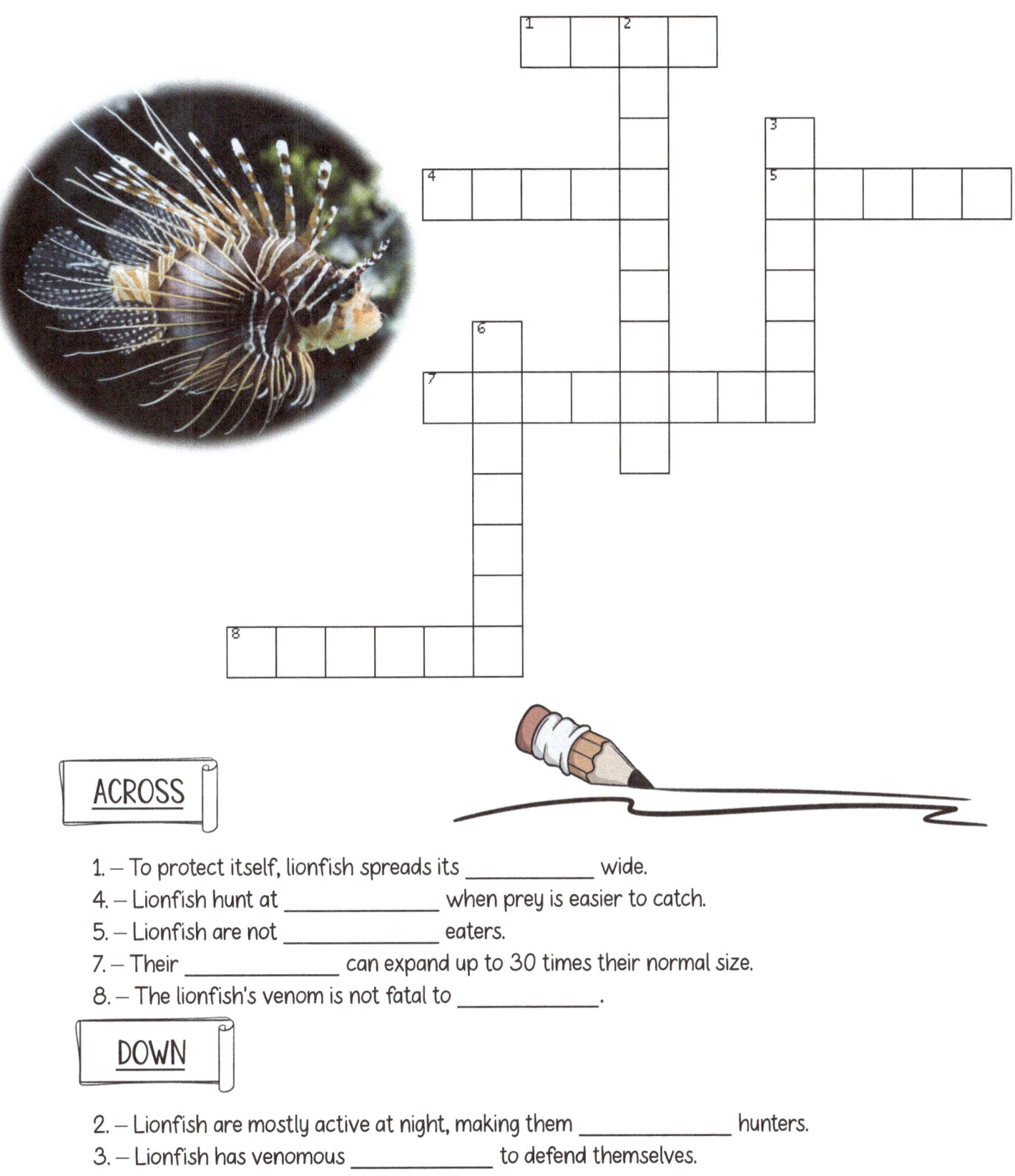

ACROSS

1. — To protect itself, lionfish spreads its _____ wide.
4. — Lionfish hunt at _____ when prey is easier to catch.
5. — Lionfish are not _____ eaters.
7. — Their _____ can expand up to 30 times their normal size.
8. — The lionfish's venom is not fatal to _____.

DOWN

2. — Lionfish are mostly active at night, making them _____ hunters.
3. — Lionfish has venomous _____ to defend themselves.
6. — Lionfish are known for their brightly colored _____.

Clownfish

ACROSS

2. — Clownfish eat _____ and plankton.

5. — The movie "Finding _____" features a clownfish.

6. — Clownfish and anemones work together in a _____ relationship.

7. — Clownfish are orange and white with black _____.

DOWN

1. — Clownfish are covered in slippery _____ to stay safe.

2. — Clownfish are also known as Clown _____.

3. — Clownfish live in coral reefs and _____.

4. — The sea _____ is the clownfish's home.

Angelfish

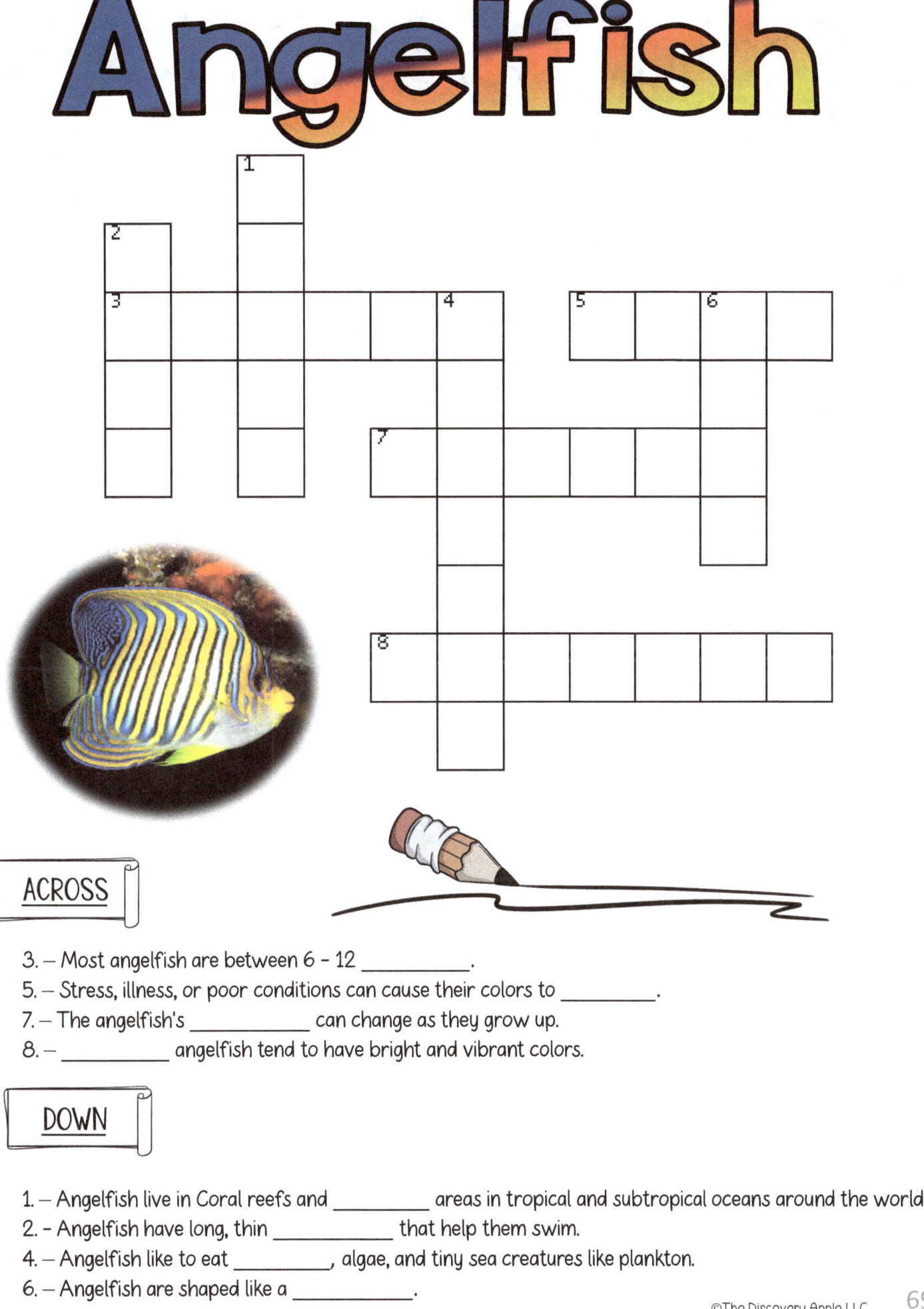

ACROSS

3. — Most angelfish are between 6 – 12 _____.
5. — Stress, illness, or poor conditions can cause their colors to _____.
7. — The angelfish's _____ can change as they grow up.
8. — _____ angelfish tend to have bright and vibrant colors.

DOWN

1. — Angelfish live in Coral reefs and _____ areas in tropical and subtropical oceans around the world.
2. — Angelfish have long, thin _____ that help them swim.
4. — Angelfish like to eat _____, algae, and tiny sea creatures like plankton.
6. — Angelfish are shaped like a _____.

Pygmy Seahorse

ACROSS

1. – The pygmy seahorse's body match the color and bumpy texture of _____.
2. – Pygmy seahorses live in coral reefs and sea fans in _____ waters of the Indo-pacific.
5. – The scientific name of the pygmy seahorse is _____ Bargibanti.
6. – The _____ pygmy seahorse is native to the waters around Japan.
7. – The _____ pygmy seahorse is the tiniest seahorse.

DOWN

1. – Pygmy seahorses are masters of _____, making them hard to find.
3. – Seahorses swim _____.
4. – The _____ pygmy seahorse is native to the waters around Japan.

Harlequin Shrimp

ACROSS

4. — Another common name for the harlequin shrimp is the _____ shrimp.
5. — Using their strong, flat _____, harlequin shrimp work together to capture sea stars many times their size.
8. — Harlequin shrimp often work in _____ with one shrimp acting as a lookout while the other eats.

DOWN

1. — They live in warm, tropical waters of the _____ and Pacific Oceans.
2. — They have long, thin _____ that help them feel and sense motion in the water.
3. — Harlequin shrimp move slowly while waving their _____ and antennae.
6. — Harlequin shrimp eat sea _____, which they flip over to get to their soft parts.
7. — They like to hide during the day and come out at night to _____.

Peacock Mantis Shrimp

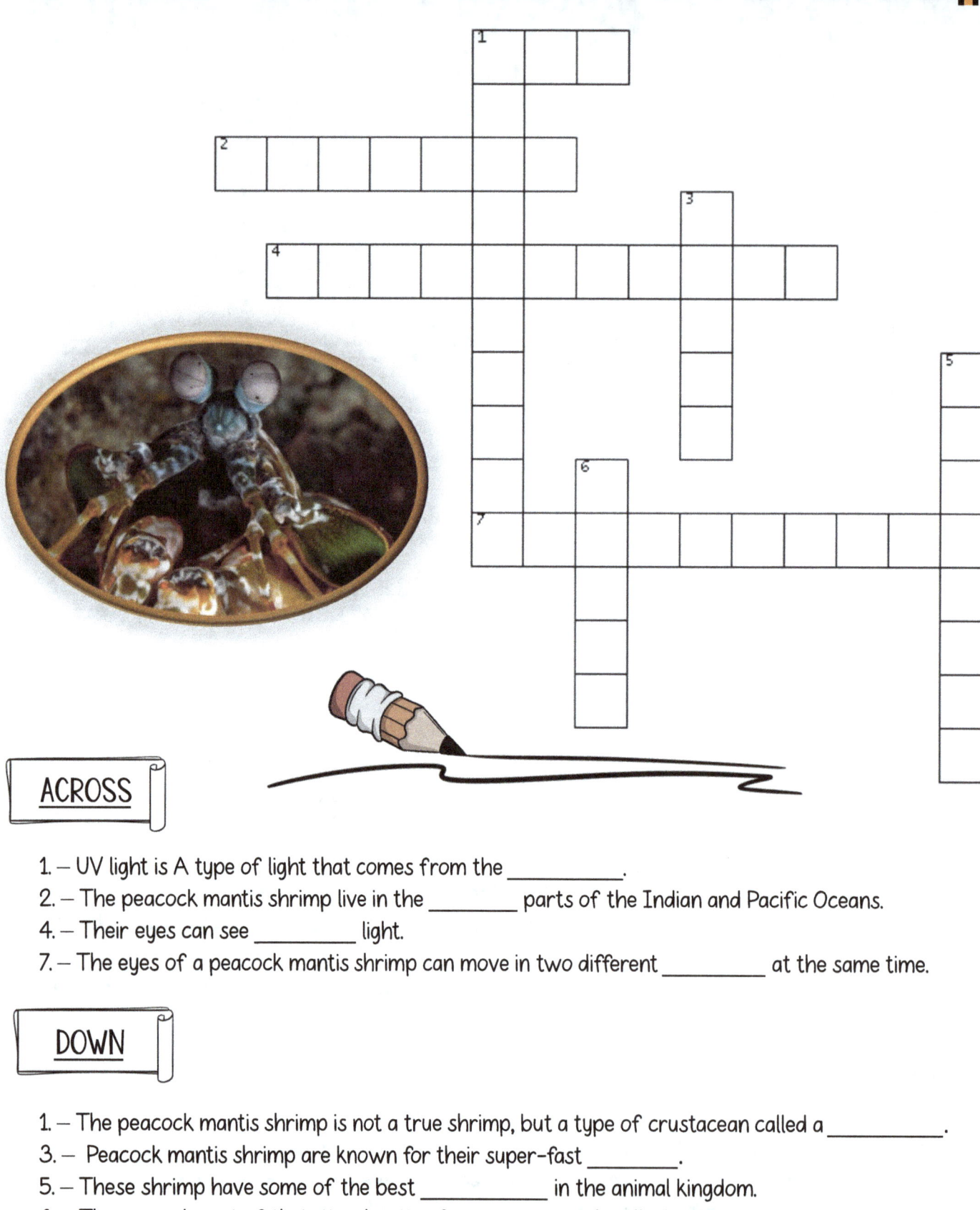

ACROSS

1. — UV light is A type of light that comes from the _____.

2. — The peacock mantis shrimp live in the _____ parts of the Indian and Pacific Oceans.

4. — Their eyes can see _____ light.

7. — The eyes of a peacock mantis shrimp can move in two different _____ at the same time.

DOWN

1. — The peacock mantis shrimp is not a true shrimp, but a type of crustacean called a _____.

3. — Peacock mantis shrimp are known for their super-fast _____.

5. — These shrimp have some of the best _____ in the animal kingdom.

6. — They spend most of their time hunting for _____ and mollusks to eat.

Spanish Shawl Sea Slug

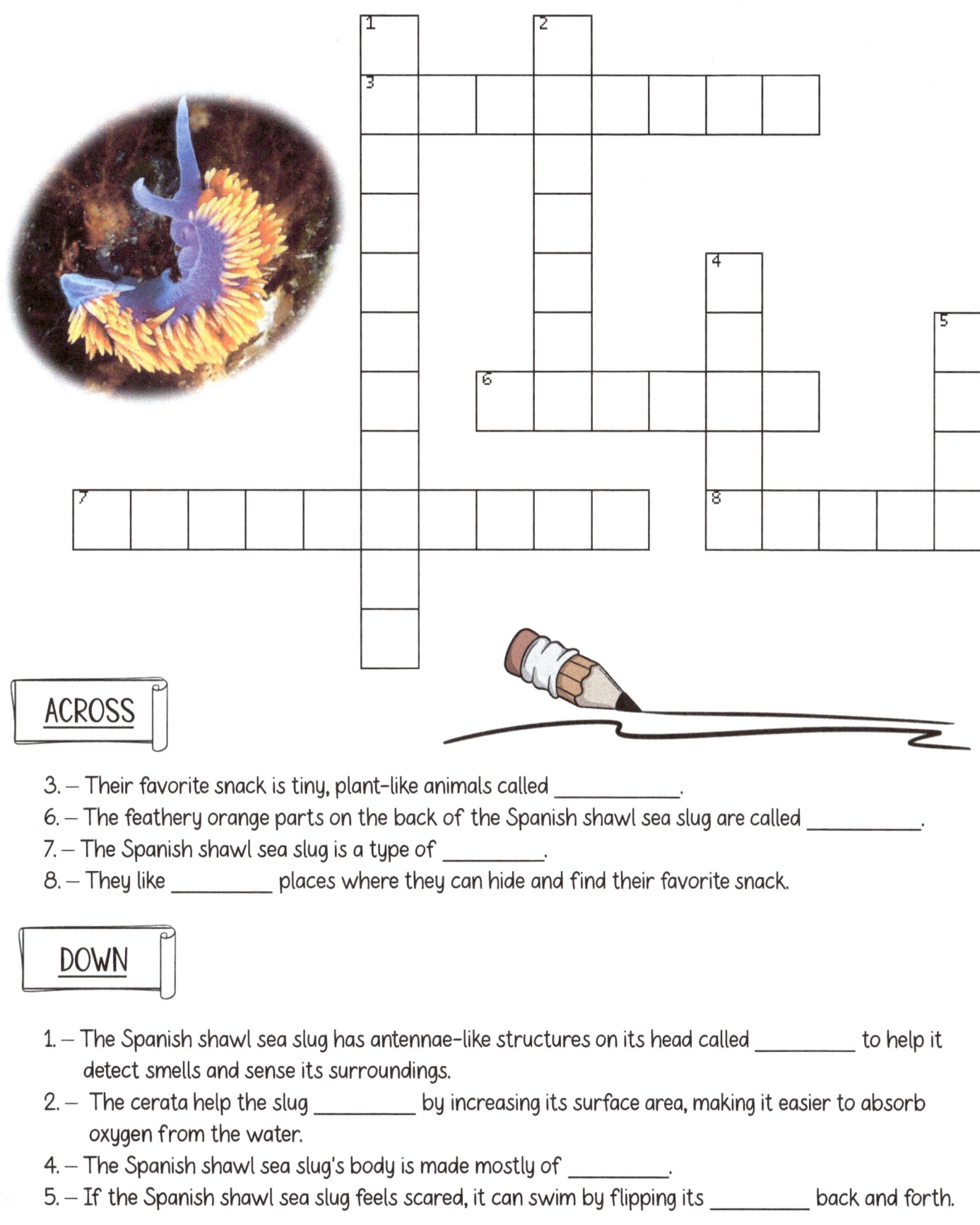

ACROSS

3. — Their favorite snack is tiny, plant-like animals called _____.
6. — The feathery orange parts on the back of the Spanish shawl sea slug are called _____.
7. — The Spanish shawl sea slug is a type of _____.
8. — They like _____ places where they can hide and find their favorite snack.

DOWN

1. — The Spanish shawl sea slug has antennae-like structures on its head called _____ to help it detect smells and sense its surroundings.
2. — The cerata help the slug _____ by increasing its surface area, making it easier to absorb oxygen from the water.
4. — The Spanish shawl sea slug's body is made mostly of _____.
5. — If the Spanish shawl sea slug feels scared, it can swim by flipping its _____ back and forth.

Leaf Sheep

3. — The leaf sheep sea slug looks like a cartoon sheep covered in _____.
4. — The feathery structures on a leaf sheep's back are called _____.
6. — Leaf sheep perform _____, a process where they use chloroplasts from algae to make energy from sunlight
8. — Leaf sheep are often called _____- powered sea slugs.

DOWN

1. — Leaf sheep can use sunlight for energy through _____.
2. — The scientific name of the leaf sheep is _____ kuroshimae.
5. — Leaf sheep get their green color from eating _____.
7. — The leaf sheep is a _____ sea slug, usually less than half an inch long.

Blue Dragon

ACROSS

4. — When blue dragons eat Portuguese man of wars, they can steal their stinging cells and store them in their finger-like _____.

7. — Blue dragons eat dangerous animals called _____ man-of-war.

DOWN

1. — The scientific name of a blue dragon is Glaucus _____.

2. — Blue dragons can store the stinging cells of their prey to _____ themselves.

3. — Blue dragons live in _____ oceans around the world.

5. — The blue dragon is often spotted _____ up-side down.

6. — The blue dragon is a type of sea _____.

Mimic Octopus

2. – The mimic octopus has the ability to change its _____.

3. – They prefer to live in places with sandy or _____ bottoms.

5. – Mimic octopuses live in the warm waters near Indonesia, Malaysia, and other parts of southeast _____.

6. – The mimic octopus can flatten its body and arms to look like a _____.

7. – It can switch between disguises in _____.

DOWN

1. – The mimic octopus can imitate other animals to avoid _____.

4. – The mimic octopus can pretend to be other _____.

8. – Mimic octopuses are active during the _____.

Blue-Ringed Octopus

1. — This octopus has _____ arms.
4. — The scientific name of the blue-ringed octopus is _____.
6. — The blue-ringed octopus flashes its blue _____ when it feels threatened.
7. — The blue-ringed octopus has the ability to change its _____ and texture to blend perfectly with its surroundings.
8. — The blue-ringed octopus has some of the most powerful _____ in all the oceans.

DOWN

2. — The blue-ringed octopus is an expert at _____.
3. — The blue-ringed octopus eats small _____, fish and crabs.
5. — The blue-ringed octopus is extremely _____, with venom that can harm humans.

77

Mauve Stinger Jellyfish

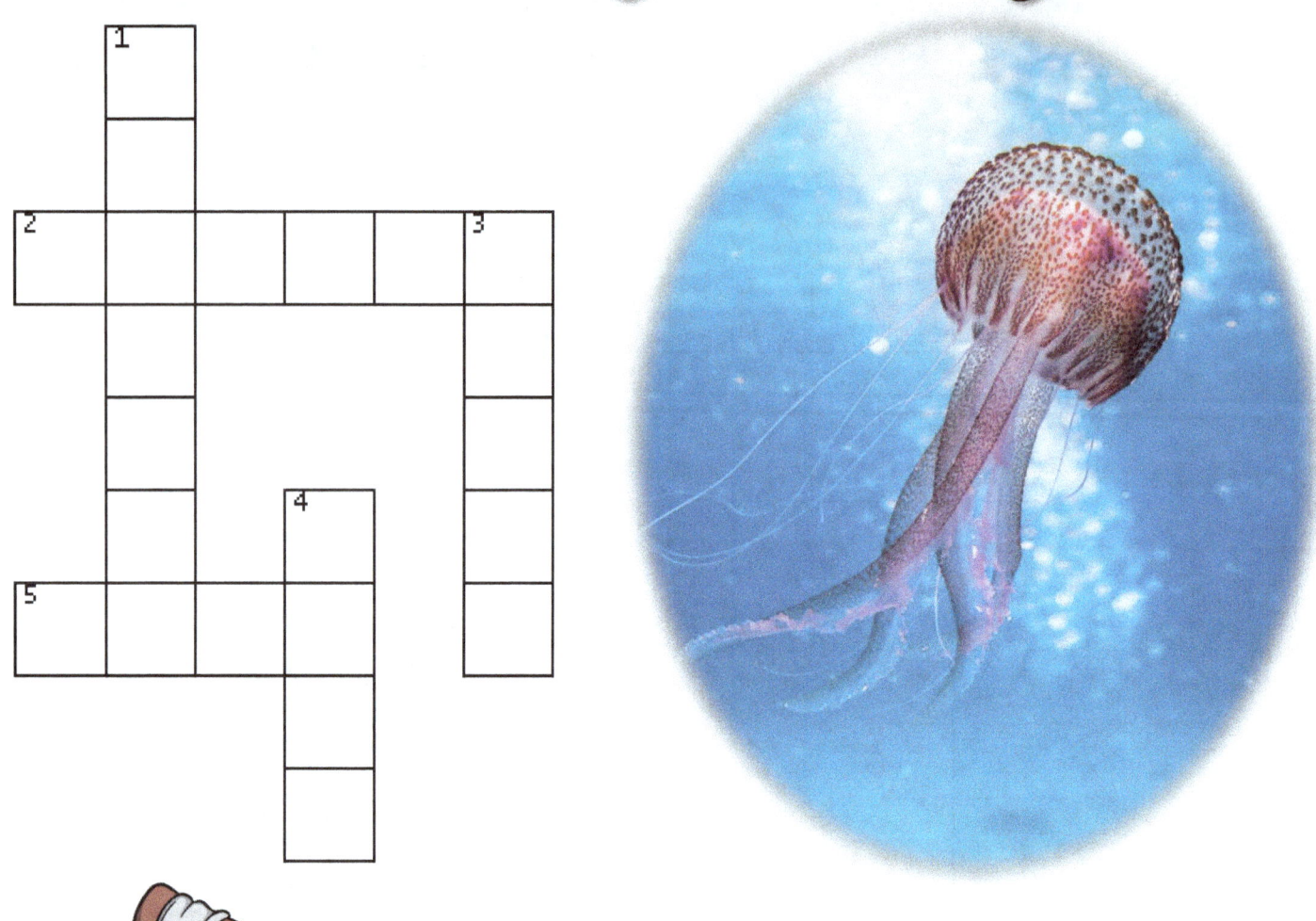

ACROSS

2. – The mauve stinger jellyfish often travels in massive groups called _____.

5. – The mauve stinger jellyfish glows in the _____.

DOWN

1. – The scientific name of the mauve stinger is _____ noctiluca.

3. – The mauve stinger has tentacles that can _____ when touched.

4. – It makes a glowing slime from its _____.

Leafy Seadragon

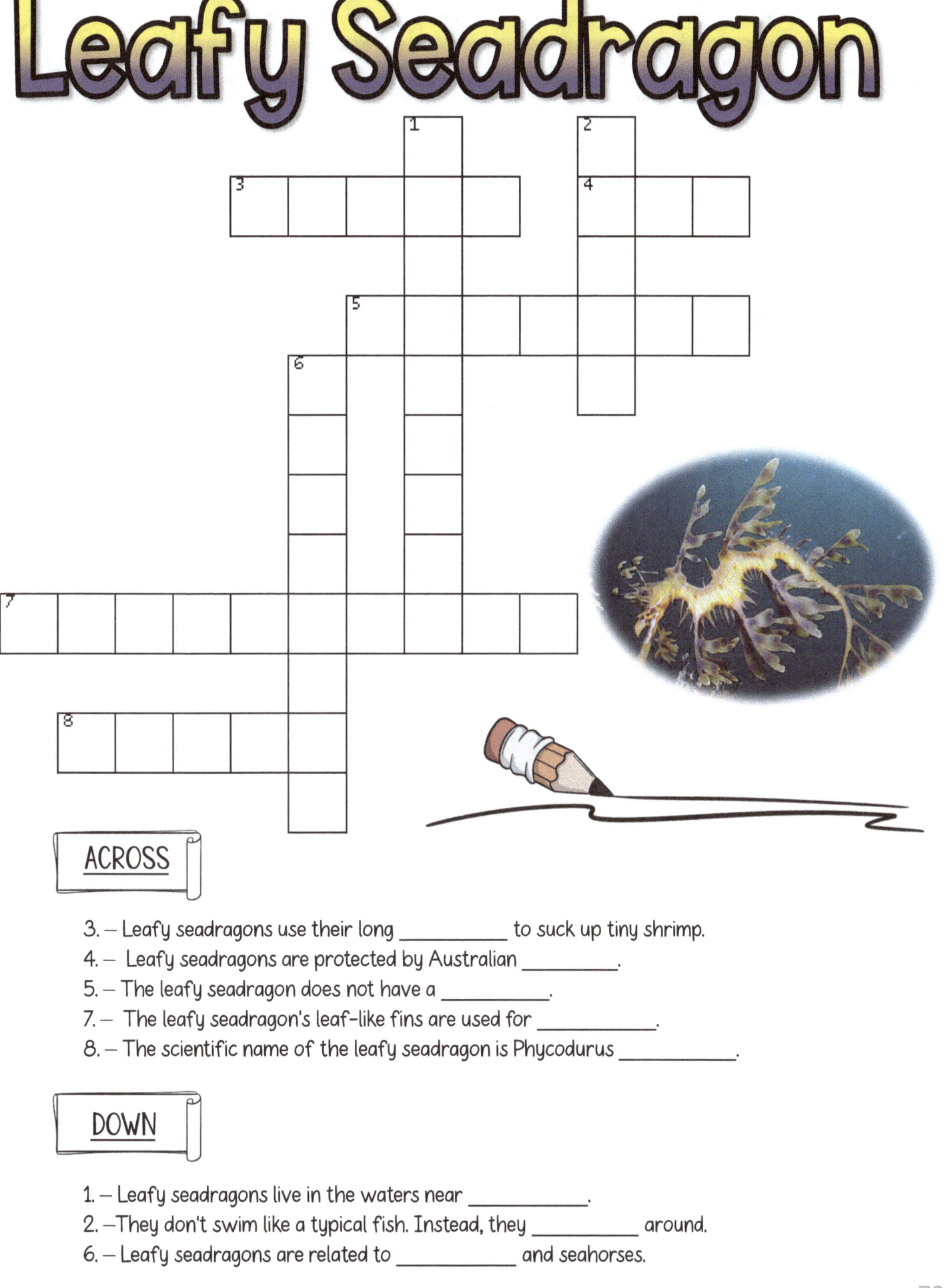

ACROSS

3. — Leafy seadragons use their long _____ to suck up tiny shrimp.

4. — Leafy seadragons are protected by Australian _____.

5. — The leafy seadragon does not have a _____.

7. — The leafy seadragon's leaf-like fins are used for _____.

8. — The scientific name of the leafy seadragon is Phycodurus _____.

DOWN

1. — Leafy seadragons live in the waters near _____.

2. — They don't swim like a typical fish. Instead, they _____ around.

6. — Leafy seadragons are related to _____ and seahorses.

Christmas Tree Worms

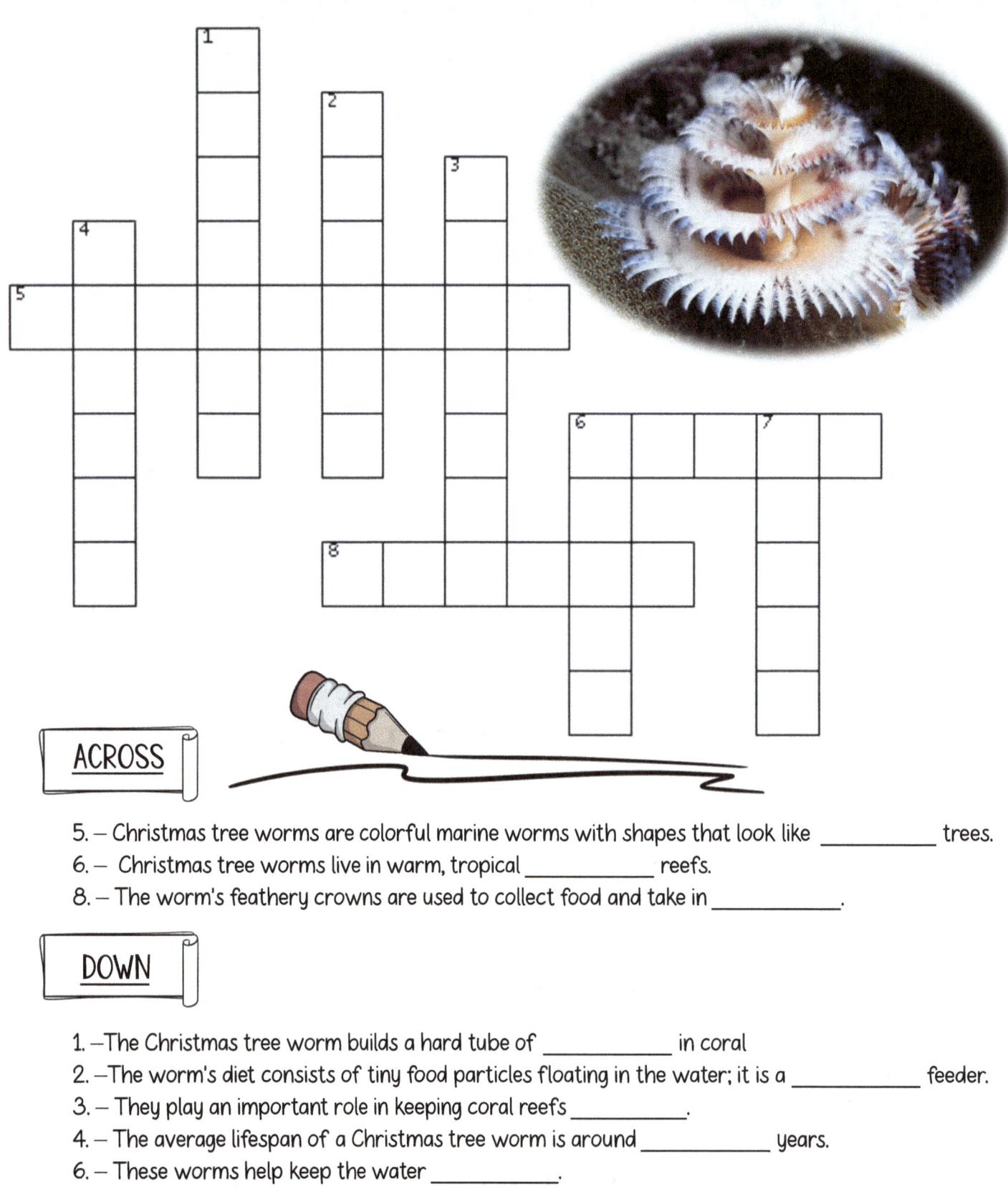

ACROSS

5. — Christmas tree worms are colorful marine worms with shapes that look like _____ trees.

6. — Christmas tree worms live in warm, tropical _____ reefs.

8. — The worm's feathery crowns are used to collect food and take in _____.

DOWN

1. —The Christmas tree worm builds a hard tube of _____ in coral

2. —The worm's diet consists of tiny food particles floating in the water; it is a _____ feeder.

3. — They play an important role in keeping coral reefs _____.

4. — The average lifespan of a Christmas tree worm is around _____ years.

6. — These worms help keep the water _____.

7. — Christmas tree worms prevent _____ from overgrowing the coral.

Sea Anemones

ACROSS

4. — Unlike plants, sea anemones don't need _____ to survive.
5. — Sea anemones use their _____ to sting and catch prey.
6. — Sea anemones are _____, which means they eat meat like small fish and plankton.
8. — They often have a _____ relationship with clownfish.

DOWN

1. — Sea anemones look like underwater _____, but they are animals.
2. — Sea anemones are close relatives of coral and _____.
3. — The scientific name for sea anemones is _____.
7. — Some species of anemones are believed to live over _____ years.

Sea Stars

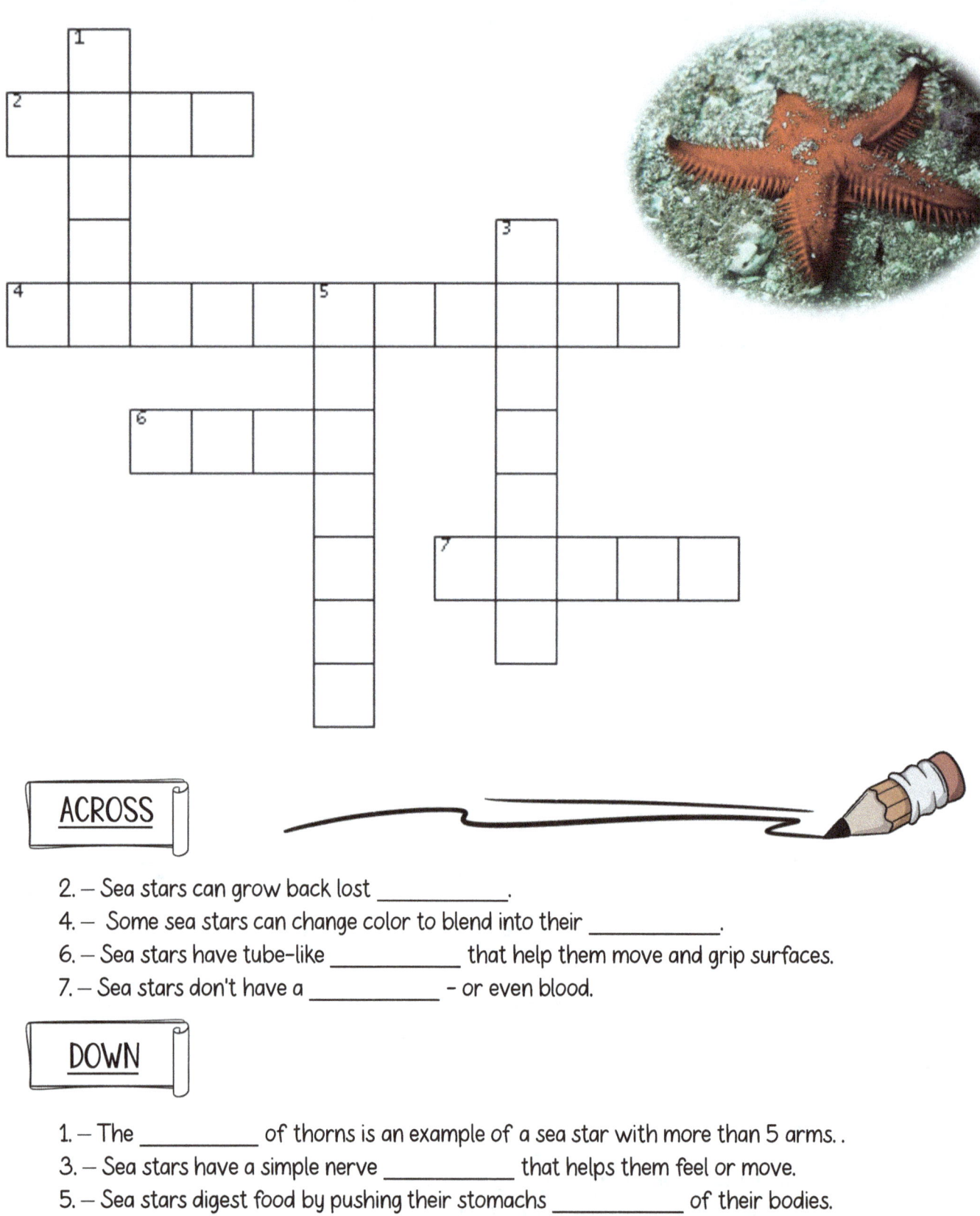

ACROSS

2. – Sea stars can grow back lost _____.

4. – Some sea stars can change color to blend into their _____.

6. – Sea stars have tube-like _____ that help them move and grip surfaces.

7. – Sea stars don't have a _____ - or even blood.

DOWN

1. – The _____ of thorns is an example of a sea star with more than 5 arms. .

3. – Sea stars have a simple nerve _____ that helps them feel or move.

5. – Sea stars digest food by pushing their stomachs _____ of their bodies.

Sea Urchins

ACROSS

4. — Sea urchins are found in every _____ on Earth.

5. — Sea urchins have sharp, spiny _____ to protect themselves.

DOWN

1. — Sea urchins are _____, meaning they eat plants like algae.

2. — Sea urchins are important for controlling the growth of _____ on coral reefs.

3. — Sea urchins can see even though they don't have _____.

ANSWER
Keys

CROSSWORD PUZZLES ANSWER KEYS

Pg 64 - Mandarinfish

Across:
- 3. OUNCE
- 5. PREDATORS
- 6. SLIMY
- 7. CORAL

Down:
- 1. SPLENDIDUS
- 2. COPEPODS
- 4. TOXIC
- 5. PACIFIC

Pg 65 - Royal Gramma Fish

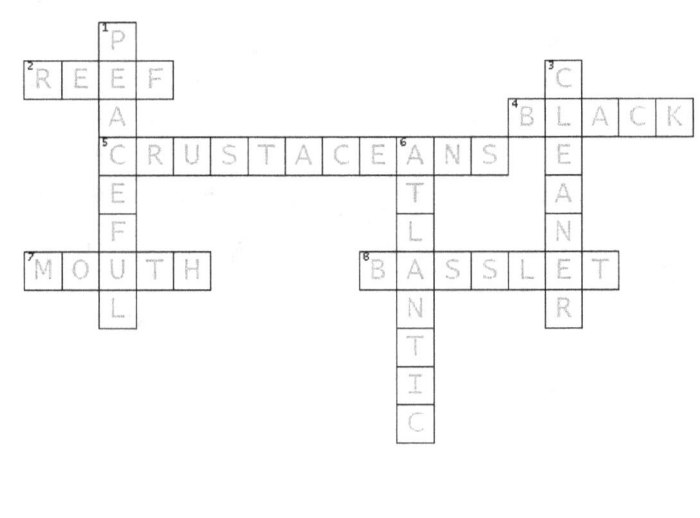

Across:
- 2. REEF
- 4. BLACK
- 5. CRUSTACEANS
- 7. MOUTH
- 8. BASSLET

Down:
- 1. PEACEFUL
- 3. CLEANER
- 4. ATLANTIC

Pg 66 - Butterflyfish

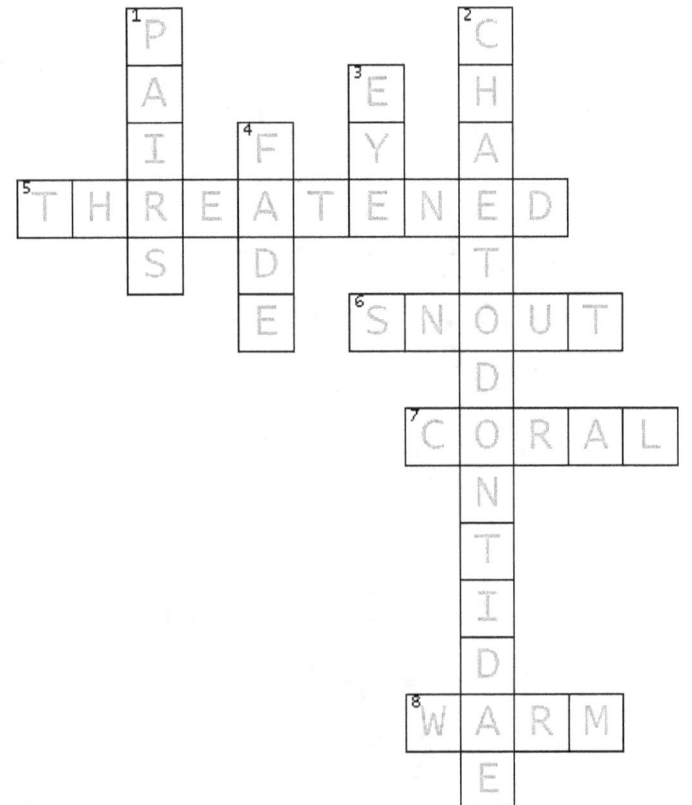

Across:
- 5. THREATENED
- 6. SNOUT
- 7. CORAL
- 8. WARM

Down:
- 1. PAIRS
- 2. CHAETODONTIDAE
- 3. EYE
- 4. FADE

Pg 67 - Lionfish

Across:
- 1. FINS
- 4. NIGHT
- 5. PICKY
- 7. STOMACHS
- 8. HUMANS

Down:
- 2. NOCTURNAL
- 3. SPINES
- 6. STRIPE

CROSSWORD PUZZLES ANSWER KEYS

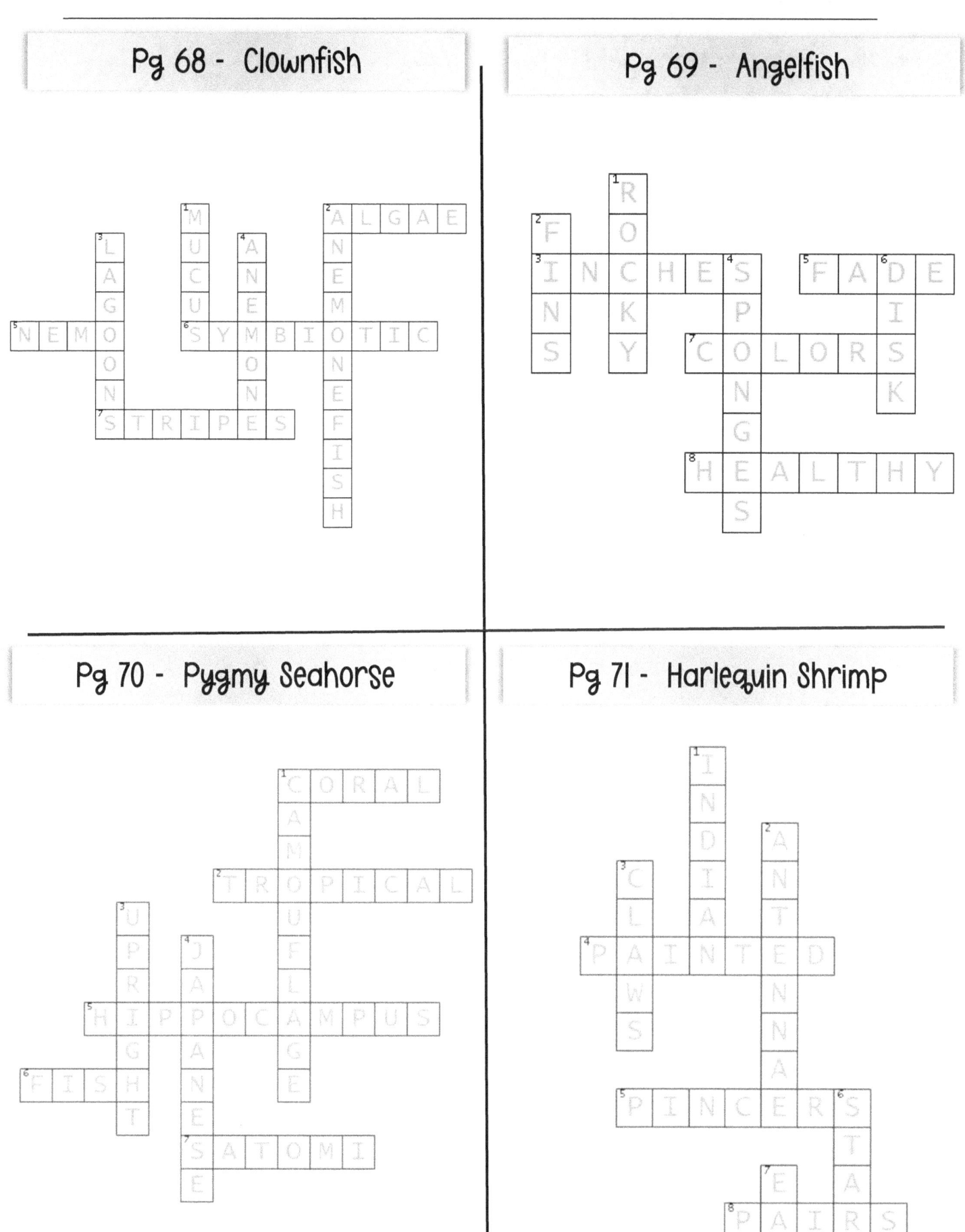

Pg 68 - Clownfish

Pg 69 - Angelfish

Pg 70 - Pygmy Seahorse

Pg 71 - Harlequin Shrimp

CROSSWORD PUZZLES ANSWER KEYS

Pg 72 - Peacock Mantis Shrimp

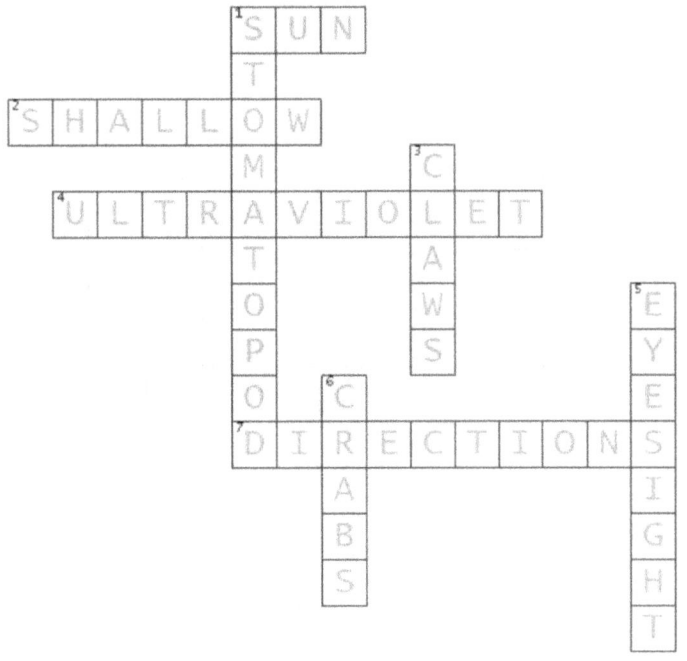

Pg 73 - Spanish Shawl Sea Slug

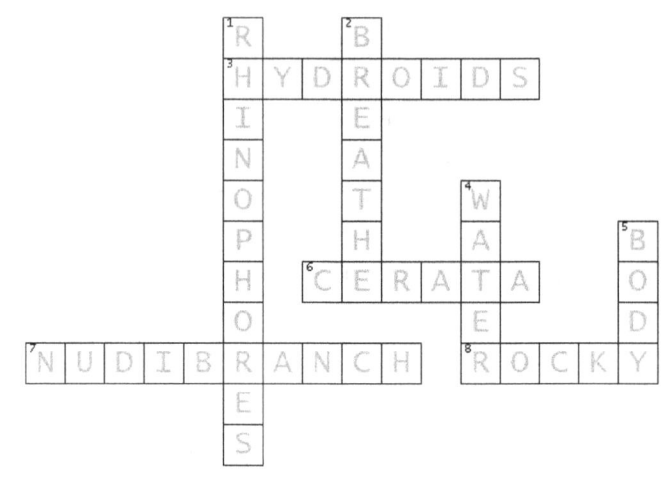

Pg 74 - Leaf Sheep

Pg 75 - Blue Dragon

CROSSWORD PUZZLES ANSWER KEYS

Pg 76 - Mimic Octopus

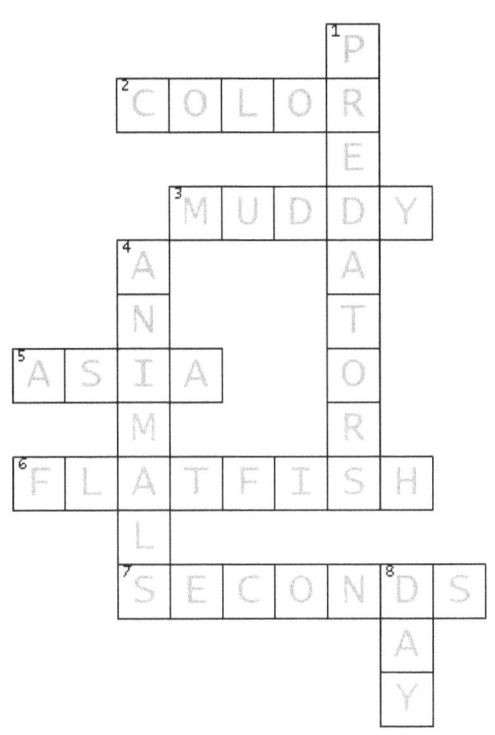

Pg 77 - Blue-ringed octopus

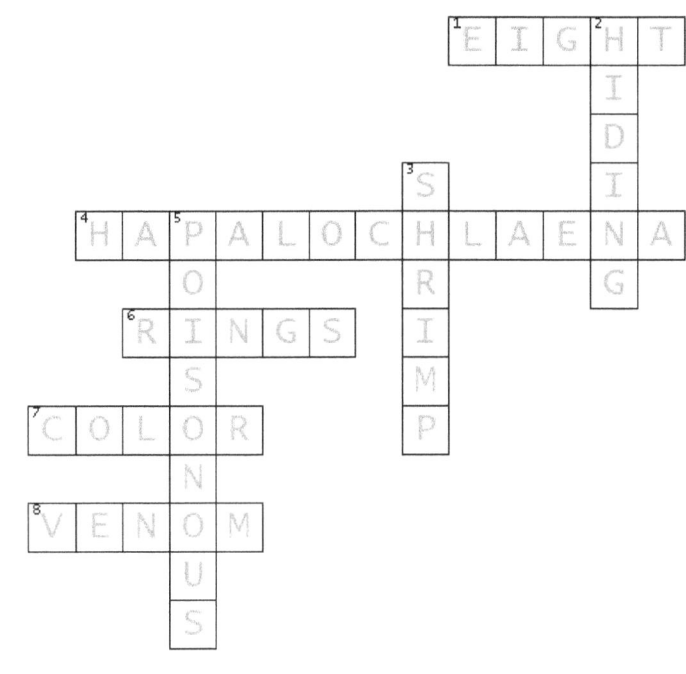

Pg 78 - Mauve Stinger Jellyfish

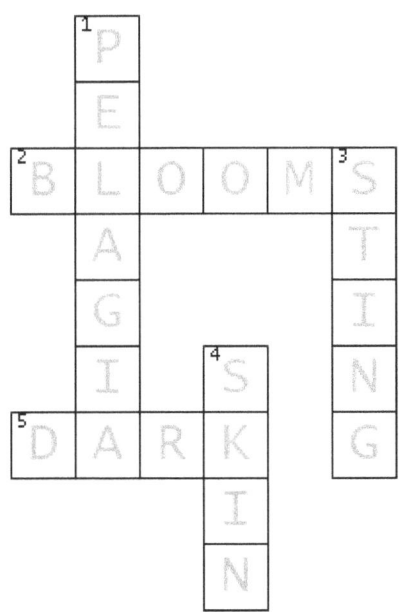

Pg 79 - Blue Dragon

CROSSWORD PUZZLES ANSWER KEYS

Pg 80 - Christmas Tree Worms

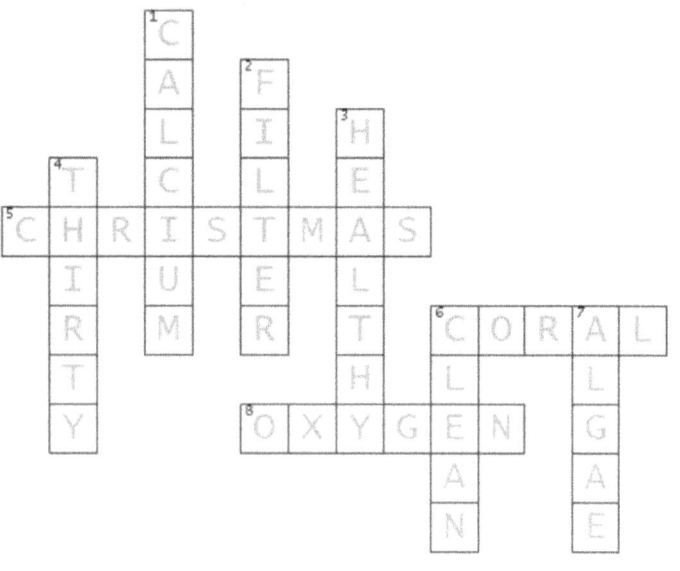

Pg 81 - Sea Anemones

Pg 82 - Sea Stars

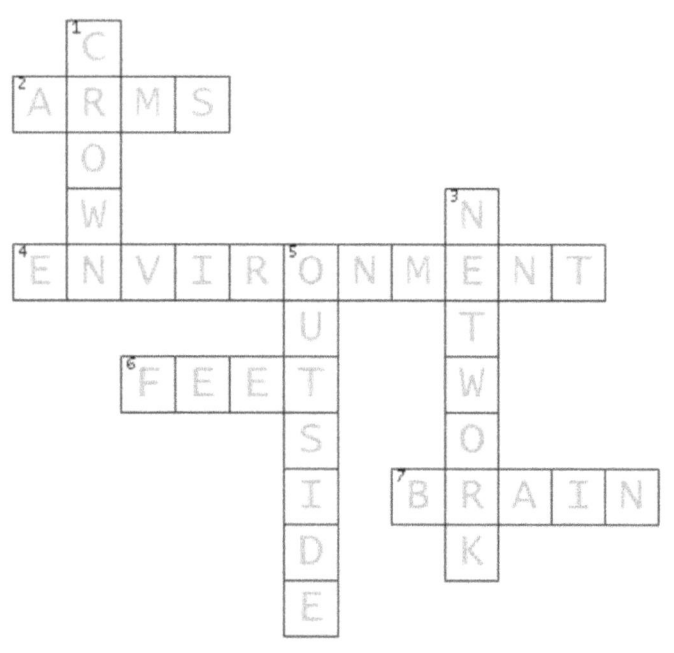

Pg 83 - Sea Urchins

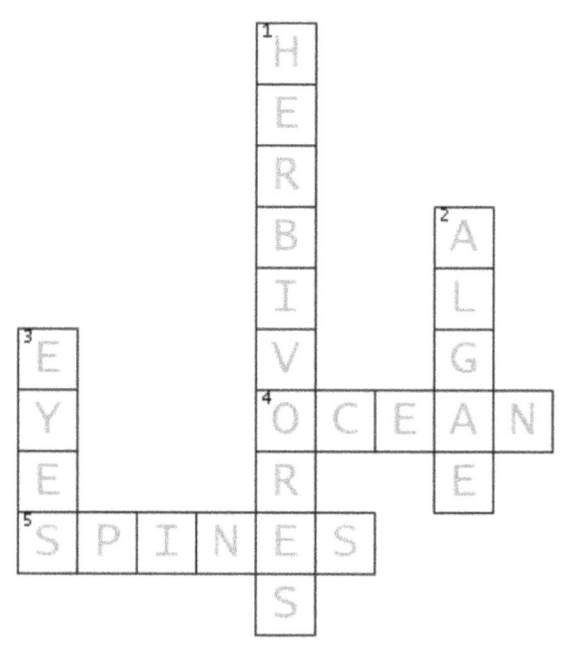

Looking for more science resources to supplement your child's learning?

Our 'Science Throughout the Year' unit provides over 1000 pages and slides of interactive activities, editable PowerPoint slides, lessons, posters, etc. (ALL DIGITAL). Nothing will be mailed to you. You would receive a zip file with all of the editable PowerPoint slides, PDFs, and Google Slides Links.

Reinforce key concepts throughout the year. Simply scan the QR code below to access the unit and enhance your child's science education.

4th – 5th Grades

What teachers are saying:

★★★★★
This resource helped me build confidence to teach science! I've always felt very intimidated and overwhelmed. These resources were so engaging! These made teaching science fun for me, which translated to engagement for my kids! I cannot say thank you enough!

- THE TRAVELING TEXAN TEACHER

★★★★★
I love this resource! The slides are beautiful and full of information. It is very well organized, the activities are engaging! The children loved all of the activities. I don't know if it had to do anything with this resource but, my students scored so much better on the FCAT than previous classes. Thank you!

- JOANN SO.

★★★★★
This was one of my best investments this year! My students loved the material and it helped make lesson planning much easier!

-SAMANTHA H.

Need science workbooks for 4ᵗʰ or 5ᵗʰ graders?

Be sure to check these out!

SCAN ME

EXPLORING LIFE **SCIENCE** IN 9 WEEKS — WORKBOOK GRADES 4-5

EXPLORING PHYSICAL **SCIENCE** IN 9 WEEKS — WORKBOOK GRADES 4-5

EXPLORING EARTH **SCIENCE** IN 9 WEEKS — WORKBOOK GRADES 4-5

EXPLORING SPACE **SCIENCE** IN 9 WEEKS — WORKBOOK GRADES 4-5

4 WORKBOOKS in **1**

EXPLORING **SCIENCE** THROUGHOUT THE YEAR — GRADES 4-5 — 4 WORKBOOKS IN 1

ENTIRE YEAR! OF SCIENCE WORK

Want Access to my FREE Resource Library?

Free Resources!

All in one place!

SCAN ME

PHOTOS - CREDITS

PHOTOS - CREDITS

THE OCEAN

Is full of secrets.

KEEP EXPLORING,

And you will always

DISCOVER

Something new!